Julian Sonksen

Sensor Process and Device for Determining a Physical Value

Julian Sonksen

Sensor Process and Device for Determining a Physical Value

DE10 2004 037 519 B4 A Proof of Concept

Südwestdeutscher Verlag für Hochschulschriften

Impressum/Imprint (nur für Deutschland/ only for Germany)
Bibliografische Information der Deutschen Nationalbibliothek: Die Deutsche Nationalbibliothek
verzeichnet diese Publikation in der Deutschen Nationalbibliografie; detaillierte bibliografische
Daten sind im Internet über http://dnb.d-nb.de abrufbar.
 Alle in diesem Buch genannten Marken und Produktnamen unterliegen warenzeichen-, marken-
oder patentrechtlichem Schutz bzw. sind Warenzeichen oder eingetragene Warenzeichen der
jeweiligen Inhaber. Die Wiedergabe von Marken, Produktnamen, Gebrauchsnamen,
Handelsnamen, Warenbezeichnungen u.s.w. in diesem Werk berechtigt auch ohne besondere
Kennzeichnung nicht zu der Annahme, dass solche Namen im Sinne der Warenzeichen- und
Markenschutzgesetzgebung als frei zu betrachten wären und daher von jedermann benutzt
werden dürften.

Verlag: Südwestdeutscher Verlag für Hochschulschriften Aktiengesellschaft & Co. KG
Dudweiler Landstr. 99, 66123 Saarbrücken, Deutschland
Telefon +49 681 37 20 271-1, Telefax +49 681 37 20 271-0
Email: info@svh-verlag.de
Zugl.: Kassel, Universität Kassel, Dissertation, 2010

Herstellung in Deutschland:
Schaltungsdienst Lange o.H.G., Berlin
Books on Demand GmbH, Norderstedt
Reha GmbH, Saarbrücken
Amazon Distribution GmbH, Leipzig
ISBN: 978-3-8381-1859-8

Imprint (only for USA, GB)
Bibliographic information published by the Deutsche Nationalbibliothek: The Deutsche
Nationalbibliothek lists this publication in the Deutsche Nationalbibliografie; detailed
bibliographic data are available in the Internet at http://dnb.d-nb.de.
 Any brand names and product names mentioned in this book are subject to trademark, brand
or patent protection and are trademarks or registered trademarks of their respective holders.
The use of brand names, product names, common names, trade names, product descriptions
etc. even without a particular marking in this works is in no way to be construed to mean that
such names may be regarded as unrestricted in respect of trademark and brand protection
legislation and could thus be used by anyone.

Publisher: Südwestdeutscher Verlag für Hochschulschriften Aktiengesellschaft & Co. KG
Dudweiler Landstr. 99, 66123 Saarbrücken, Germany
Phone +49 681 37 20 271-1, Fax +49 681 37 20 271-0
Email: info@svh-verlag.de

Printed in the U.S.A.
Printed in the U.K. by (see last page)
ISBN: 978-3-8381-1859-8

Copyright © 2010 by the author and Südwestdeutscher Verlag für Hochschulschriften
Aktiengesellschaft & Co. KG and licensors
All rights reserved. Saarbrücken 2010

Title: DE 10 2004 037 519 B4 "Sensor Process and Device for Determining a Physical Value" - A Proof of Concept

Author: Julian H. B. Sonksen

Supervisors: Prof. Dr. rer. nat. habil. H. Hillmer / Prof. Dr.-Ing. habil. J. Börcsök

Abstract: A two mode external cavity semiconductor laser system is investigated to show the validity of the ideas and concepts fundamental to patent DE 10 2004 037 519 B4. The development of an existing laser to the functional sensor device is documented and its optimization explained. The final system is characterized in detail. The actual proof of concept consist of the investigation of the laser system's output to the introduction of a selective absorber into the external laser cavity. As a result, linear dependence of mode intensity ratio on absorber concentration is obtained. Furthermore, the sensitivity of the system can be increased by decreasing the mode spacing of the two modes of the laser.

Key Words: Laser, External Cavity, ICAS, Fiber Optics

Date of Disputation: June 1st, 2010

Titel: DE 10 2004 037 519 B4 "Sensor-Vorrichtung und Verfahren zur Ermittlung einer physikalischen Größe" - Ein Proof of Concept

Autor: Julian H. B. Sonksen

Referenten: Prof. Dr. rer. nat. habil. H. Hillmer / Prof. Dr.-Ing. habil. J. Börcsök

Zusammenfassung: Ein Diodenlaser mit externer Kavität wird untersucht, um die dem Patent DE 10 2004 037 519 B4 zugrundeliegenden Ideen und Konzepte auf Gültigkeit zu überprüfen. Hierzu werden die Entwicklung eines existierenden Lasers zu einem funktionierenden Sensor und seine Optimierung dokumentiert. Das finale Lasersystem wird im Detail charakterisiert. Der eigentliche Proof of Concept besteht schließlich aus der Evaluation der Antwort des Lasersystems auf die Einbringung eines selektiven Absorbers in die Laserkavität. Als Ergebnis steht eine lineare Abhängigkeit des Intensitätsverhältnisses der beiden im System anschwingenden Moden von der Konzentration des Absorbers. Weiterhin wird gezeigt, dass die Empfindlichkeit mit reduziertem Modenabstand zunimmt.

Stichwörter: Laser, Externe Kavität, ICAS, Faseroptik

Datum der Disputation: 1. Juni 2010

Digest

At present, a fraction of 0.1 - 0.2% of the patients undergoing surgery become aware during the process. The situation is referred to as anesthesia awareness and is obviously very traumatic for the person experiencing it. The reason for its occurrence is mostly an insufficient dosage of the narcotic Propofol combined with the incapability of the technology monitoring the depth of the patient's anesthetic state to notice the patient becoming aware.

A solution can be a highly sensitive and selective real time monitoring device for Propofol based on optical absorption spectroscopy. Its working principle has been postulated by Prof. Dr. habil. H. Hillmer and formulated in DE 10 2004 037 519 B4, filed on Aug 30th, 2004. It consists of the exploitation of Intra Cavity Absorption effects in a two mode laser system.

In this Dissertation, a two mode external cavity semiconductor laser, which has been developed previously to this work is enhanced and optimized to a functional sensor. Enhancements include the implementation of variable couplers into the system and the implementation of a collimator arrangement into which samples can be introduced. A sample holder and cells are developed and characterized with a focus on compatibility with the measurement approach. Further optimization concerns the overall performance of the system: scattering sources are reduced by re-splicing all fiber-to-fiber connections, parasitic cavities are eliminated by suppressing the Fresnel reflexes of all one fiber ends by means of optical isolators and wavelength stability of the system is improved by the implementation of thermal insulation to the Fiber Bragg Gratings (FBG).

The final laser sensor is characterized in detail thermally and optically. Two separate modes are obtained at 1542.0 and 1542.5 nm, tunable in a range of 1 nm each. Mode Full Width at Half Maximum (FWHM) is 0.06 nm and Signal to Noise Ratio (SNR) is as high as 55 dB. Independent of tuning the two modes of the system can always be equalized in intensity, which is important as the delicacy of the intensity equilibrium is one of the main sensitivity enhancing effects formulated in DE 10 2004 037 519 B4.

For the proof of concept (POC) measurements the target substance Propofol is diluted in the solvents Acetone and DiChloroMethane (DCM), which have been investigated for compatibility with Propofol beforehand. Eight measurement series (two solvents, two cell lengths and two different mode spacings) are taken, which draw a uniform picture: mode intensity ratio responds linearly to an increase of Propofol in all cases. The slope of the linear response indicates the sensitivity of the system. The eight series are split up into two groups: measurements taken in long cells and measurements taken in short cells. Elimination of cell length from these measurements provides a clear re-grouping of the measurement series indicating a higher sensitivity at lower mode spacing. A conservative estimate of the system's sensitivity limit shows its performance on eye level with recent state of the art single mode ICAS systems, even in this very first approach.

As a conclusion it can be stated that the sensor responds reproducibly to Propofol. This response is linear. From the increased sensitivity at smaller mode spacing we can conclude that mode competition exists. Hence, the proof of concept attempted in this work is regarded as a success.

Contents

1. **Introduction**   1
   - 1.1. Improving the Quality of Anesthesia   1
   - 1.2. State of Research on Competing Sensor Technology   2
   - 1.3. Legal Framework   4
   - 1.4. Objective and Chapter Outline   6

2. **Fundamentals leading to an ICAS Experiment**   9
   - 2.1. The General Principle of Lasing   10
     - 2.1.1. Transitions in the Active Semiconductor Material   10
     - 2.1.2. Inversion as the prerequisite for Lasing   12
     - 2.1.3. Mathematical Description of the Lasing Process   12
     - 2.1.4. The Effect of the Resonator on Laser Emission   14
     - 2.1.5. The Threshold Condition   16
     - 2.1.6. Parameters Frequently Used in Laser Characterization   18
   - 2.2. Resonator Designs Interesting for ICAS   19
     - 2.2.1. The Fabry-Pérot Resonator   19
     - 2.2.2. Exploiting Bragg Reflection for Laser Resonators   23
     - 2.2.3. Tailoring Laser Emission with External Cavities   25
   - 2.3. Intra Cavity Absorption Spectroscopy in the NIR   27
     - 2.3.1. EM-Active Transitions in NIR: Rotational-Vibrational Spectra   27
     - 2.3.2. Relevant Sources of Absorption Line Broadening   33
     - 2.3.3. Intra Cavity Absorption   37
     - 2.3.4. Sensitivity Limitations of ICAS   39
     - 2.3.5. The Advantages of Two Mode ICAS   41

3. **Laser Setup Evolution**   47
   - 3.1. Where this work begins: The Status Quo   47
     - 3.1.1. Laser System Presentation   47

Contents

      3.1.2. Analyzing System Emission for Suitability in ICAS 48
      3.1.3. Introduction of the Wavelength Tuning Mechanism 50
  3.2. Introducing the Modifcations to Obtain a Functional ICAS Setup 52
      3.2.1. Equalizing Intensities and Tailoring Threshold 52
      3.2.2. The Necessity for a Freely Propagating Beam 57
      3.2.3. Thermal Instability in the FBG Feedback System 62
      3.2.4. The Elimination of Parasitic Cavities 65
      3.2.5. Reducing Linewidth 70
  3.3. Liquid Cell Implementation 72
      3.3.1. Optical Prerequisites of the Cells Used 72
      3.3.2. Cell Material and Design 73
      3.3.3. The Design of a Sample Cell Holder 75
  3.4. Final Setup for the Proof of Concept 78

4. Experimental Work   **81**
  4.1. Characterization of the Enhanced Laser System 81
      4.1.1. Analyzing Laser Output 81
      4.1.2. The Effect of Temperature and Injection Current on the Gain Profile . . 83
      4.1.3. Methods of Manipulating Threshold Current 87
      4.1.4. Spectral Characterization of the Laser Modes 89
      4.1.5. Wavelength Tuning Response 92
      4.1.6. Measurements on Improved Wavelength Stability 96
      4.1.7. Consequences for the POC 97
  4.2. Chemical and Optical Characterization of Propofol and Solvents 102
      4.2.1. What is Propofol? 102
      4.2.2. Evaluation of Different Solvents for Compatibility with Propofol 105
      4.2.3. How Different Solvents Bias POC Results 107

5. Proof of Concept   **109**
  5.1. Data Acquisition 109
      5.1.1. System Initialization 109
      5.1.2. Sample and Cell Preparation 110
      5.1.3. Taking Measurements 110
      5.1.4. Measurement Overview 111
  5.2. Data Evaluation 113
      5.2.1. Unmodified Output Spectra 113

	5.2.2. Comparing Integral Ratio and Maximum Intensity Ratio	114
	5.2.3. Integral Ratio versus Concentration	117
	5.2.4. Integral Ratio versus Single Pass Absorption	120
	5.2.5. Estimate of Sensitivity Limit	122
5.3.	POC Summary	124

6. Conclusion and Outlook   125

A. Appendix   141

Nomenclature

α	Absorption Coefficient
β	Filling Factor
χ	Molar Fraction
Δ	Indicates a Difference
δ	Indicates Thermal Fluctuation
ϵ	Dielectric Constant
η_{nat}	Normalization Factor
Γ	Confinement Factor
γ	Damping Factor
$\hbar$	Planck Constant divided by 2π
κ	Hook's Constant
λ	Wavelength
$\lambda_{\text{i,o}}$	Wavelength of the Inner and Outer Fiber Bragg Grating
$\text{FT}\left(x\left(t\right)\right)$	Fourier Transformation
$\mathscr{A}$	Einstein Coefficient for Spontaneous Emission
$\mathscr{B}_{\text{mn}}$	Einstein Coefficient for Absorption
$\mathscr{B}_{\text{nm}}$	Einstein Coefficient for Stimulated Emission
$\mathscr{E}$	Constants Introduced for Simplification
$\mathscr{F}$	Finesse
$\mathscr{I}$	Intensity
$\mathscr{N}$	Number of Molecules per Unit Volume
ν	Frequency of Incident Radiation
Ω	Anharmonicity of the Morse Potential
ω	Angular Frequency
Φ	Number of Photons
ϕ	Gaussian Error
σ	Collision Cross Section
τ	Life Time

Contents

Θ	Temperature
θ	Moment of Inertia
$\tilde{\nu}$	Wavenumber
Υ	Mode Intensity Ratio
ς_b	Factor for Line Width Broadening
ς_s	Factor for Line Shift
Ξ	Isolation
ξ	Number of Moles of Substance
ζ	Collision Frequency
A	Absorption
B	Rotation Constant
c	Velocity of Light
D	Independent Function
d	Collision Diameter
E	Energy
f_i	Fitting Parameters
G	Gain
h	Planck's Constant
J	Quantum Number
j	Order of Vibration
k_B	Boltzmann Constant
L	Loss
l	Length
M	Molarity
$m_{\text{red.}}$	Reduced Mass
n	Refractive Index
N_i	Population of an Energy Level
P	Pump Rate
p	Numbering of Layers
Q	Quality factor
R	Reflectivity
S	Scattering
T	Transmission
t_{sat}	Spectral Saturation Time
V	Volume

v	Velocity
x	Oscillation Amplitude
X_q	Distortion Parameter
y_0	Fitting Parameter
AR	Anti Reflection
ASE	Amplified Spontaneous Emission
$C_{i,o}$	Inner and Outer Cavity
CRDS	Cavity Ring Down Spectroscopy
DBR	Distributed Bragg Reflector
DFB	Distributed Feedback
e^-	Electron
EM	Electro Magnetic
FBG	Fiber Bragg Grating
$FBG_{i,o}$	Inner and Outer Fiber Bragg Grating
FP	Fabry-Pérot
FSR	Free Spectral Range
FTIRS	Fourier Transform Infra Red Spectroscopy
FWHM	Full Width at Half Maximum
InGaAs	Indium Gallium Arsenide
$M_{i,o}$	Mode Belonging to the Inner and Outer Cavity
OSA	Optical Spectrum Analyzer
PAS	Photo Acoustic Spectroscopy
PID	Proportional Integral Differential
PIN	p-/i-/n-doped
ppm	Parts Per Million
SMF	Single Mode Fiber
SNR	Signal to Noise Ratio
SOA	Semiconductor Optical Amplifier
TDLAS	Tunable Diode Laser Absorption Spectroscopy
UV	Ultra Violet
$V_{i,o}$	Inner and Outer Variable Coupler

1. Introduction

1.1. Improving the Quality of Anesthesia

"Quality is never an accident; it is always the result of high intention, sincere effort, intelligent direction and skillful execution; it represents the wise choice of many alternatives."
- *William A. Foster (1917-1945)*

The quality of a product is one of main selling arguments these days. The label 'high quality' means that a product is functional, reliable and well capable of fulfilling its task. The more delicate and critical such a task is, the more effort must be put into all creation steps of a product, in order to obtain the quality required.

This concept applies across the board. In research around the world scientists are pushing the scientific frontier that little bit further by precisely adjusting each and every parameter in processes to perfection, by eliminating distorting influences from them and by closely monitoring each and every parameter of their experiments. After success in the laboratories, the processes developed in the scientific environment are eventually transferred into industry and from then on exist in every day life. All the effort in monitoring and all the precision must reliably transfer into the industrial sector as well, if high quality is to be maintained.

In the consumer market the purchase of a product is an expensive luxury [83] that only few people can afford or benefit from. In the medical sector, however, state of the art quality should not come at a price that is affordable by few - it should be the status quo for everybody, regardless of financial power. In a situation, where patient turnover and especially cost are dominating factors in medical decisions [2], the quality of the services provided sometimes suffers, especially when time pressured staff and outdated technology comes into play. A situation that shows the consequences is found in the operation room of hospitals daily. When surgery is performed, the decision of exactly how much of the narcotic Propofol [12], a commonly used narcotic introduced to the market in 1986, a patient needs in order to sleep deep enough for painless surgery without waking up while keeping anesthesia as shallow as possible to ensure a

1. Introduction

high patient turnover at the same time is a delicate decision that is taken by the anesthetist and merely bases on personal judgement and experience. As long as such a decision is made biased by time and money wrong doses resulting from human error cannot be avoided - anesthesia awareness [6], whose long term effects on the patients psyche are unknown, or death [104] can be the result.

1.2. State of Research on Competing Sensor Technology

At the moment there is hardly any reliable technology available to determine the amount of Propofol needed by a patient, let alone technology that enables continuous in-situ monitoring of how much of it is actually metabolized [45]. Contributing to the resolution of this problematic situation by enhancing the capabilities of the technology in use with close attention on the cost of product and ease of use is the ultimate motivation of this work, which deals with the development of a sensor system targeted at the detection of Propofol. The immediate task at hand is quantifying the concentration of the substance in a composition of fluids, which can be a gas like exhaled breath or liquid like blood or serum. This is difficult for two reasons. First, high sensitivity is required as the concentration of Propofol in the exhaled breath or blood is very low and secondly high selectivity must be ensured to avoid false alarms. A low cost system basing on commercial components is called for that can guarantee both properties at the same time.

Most low cost fluid sensors commercially used for different tasks in industry at present are electrochemical sensors [75] and base on the chemical properties of their respective target substances. The selective chemical reaction of the target substance with the sensor causes a change in its electrical properties. The main problem with these sensors is their immediate exposure to the fluid to be analyzed, which results in aging and puts them at risk of sensor poisoning [71, 21, 1], which cannot be tolerated in such a critical environment as the medical sector.

The poisoning effect, which ultimately limits the sensor's lifetime and reliability, is absent with optical sensors that base on absorption spectroscopy. They can be designed so their functional components are not directly exposed to the substance to be analyzed. Optical sensors obtain their selectivity from the characteristic absorption spectrum of a target substance, the so called 'fingerprint'. The conventional approach to optical measurements is to implement a broad band light source (Hg- or Xe-Lamps), pass it through a sample and spectrally record the light transmitted. As the sample reduces specific lines in the spectrum according to its fingerprint, the comparison of the incident light with the light passed through the sample yields information

1.2. State of Research on Competing Sensor Technology

on its absorption characteristics.

The application of tunable laser diodes to spectroscopy has triggered the development of a method named Tunable Diode Laser Absorption Spectroscopy (TDLAS) which exploits tunable narrow band laser emission for wavelength selection and thereby increases the sensitivity and the selectivity of single pass spectroscopic setups [40, 86]. Apart from the properties of the light source, sensitivity in single pass setups is mainly dependent on the length of the optical path through the sample, which can be increased by folding it in multi reflection cells [101, 70, 69]. Sensitivity is quantified by the effective absorption path length that accounts for the physical absorption path and its enhancement from multiple passes.

High resolution is achieved by application of Fourier methods such as in Fourier Transform Infra Red Spectroscopy (FTIRS) [87]. It bases on the evaluation of the Fourier transformation of interferograms of the light passed through a sample. TDLAS and FTIRS are applied for in-situ process monitoring in industry [59] and analysis of athmospherical gases [97, 51]. The longest path lengths implemented range around one hundred meters [97].

Modulation of the laser output improves Signal to Noise Ratio in case of TDLAS; in the special case of modulation frequencies in the audible range, however, a light induced pressure variation of gases inside the resonator is observed from the photo acoustic effect [13]. Its exploitation for spectroscopy, Photo Acoustic Spectroscopy (PAS) requires high power lasers such as CO_2 or Er^{3+} doped fiber lasers [35, 15], because the absorption signal is directly proportionate to the power of the laser [95]. It is not suited, however, for gases at very low pressures, as low pressures limit the propagation of sound. The maximum effective absorption path length achieved in with a CO_2 laser is in the range of a few km [17].

Cavity Ring Down Spectroscopy (CRDS) bases on the pulsed operation of lasers with resonators of high quality [14]. The ring down time of such a laser pulse in a passive external cavity of high quality strongly depends on the loss inside the resonator. It decreases for increasing loss introduced by an absorber in the cavity and is an indicator for absorption strength. Incremental tuning of laser wavelength delivers the full absorption spectrum of a sample. With CRDS maximum effective absorption path lengths of up to 70 km have been achieved [72].

One of the most sensitive approaches for absorption measurements is a method in which the absorber is placed inside the cavity of a multi mode laser. Such a laser reacts very delicately to narrow band loss inside its resonator; the exploitation of this effect is called Intra Cavity Absorption Spectroscopy (ICAS) and profits from increased sensitivity from mode competition and amplification by the lasing process. Effective absorption path lengths of up to 70.000 km

1. Introduction

are reported with dye lasers [8]. Diode lasers are also capable of ICAS, but inhibit a maximum a sensitivity of about 2.5 km effective absorption path length [8]. However, they are much more cost effective compared to dye lasers, which enables their implementation in cost effective and compact commercial devices [7].

1.3. Legal Framework

Patent DE 10 2004 037 519 B4

In this field of low cost intra cavity absorption spectroscopy patent DE 10 2004 037 519 B4 on an invention in the field of ICAS by Prof. Dr. H. Hillmer has been filed on Aug. 30th, 2004, by the University of Kassel. Its objective is the measurement of a physical parameter based on its manipulation of a laser system's emission. The laser system used must be capable of emitting at least two modes above threshold. The physical parameter is measured by evaluating the change in the two modes emitted under its influence. The patent was granted on March 23rd, 2006.

Compared to state of the art systems using only one laser mode or very many, the main idea described in DE 10 2004 037 519 B4 is the exploitation of the mode competition observed, when two laser modes compete for the same energy in the active zone. It particularly takes advantage of the fact that an artificially created intensity equilibrium between the two modes is very sensitive to a variety of smallest external influences. This can be a change of amplitude, intensity, spectral position or distribution of the laser's modes. As a result, a sensing device described in DE 10 2004 037 519 B4 is capable of measuring all physical parameters which directly or indirectly influence the lasers emission characteristics.

Direct parameters are temperature, injection current, geometric and optical conditions of the resonator system, whereas indirect measurements can be taken on virtually any parameter that influences the direct ones, such as: refractive index, electrical field, magnetic field, pressure, and many more. Hence, the sensor system proposed in DE 10 2004 037 519 B4 is in principle a very versatile system to sensitively measure a variety of physical parameters.

1.3. Legal Framework

Patent exploitation

For commercial exploitation of the patent DE 10 2004 037 519 B4, which was awarded the GiNo[1] Innovation Award in 2005, the University of Kassel granted an exclusive license on the patent to the inventor, Prof. Dr. H. Hillmer. This exclusive license has then been incorporated into a spin off company, Pneumolab Gbr., founded by H. Hillmer, H. Krause and V. Viereck in 2006 and co-funded by the pre-formation sponsorship EXIST-SEED[2]. In 2007, Pneumolab Gbr. was terminated and re-established under the name Pinpoint Detection Gbr. replacing H. Krause by myself, J. Sonksen, shortly after.

Since then, Pinpoint Detection Gbr. has been developing the sensor system based on the patent with a special focus on applications in medicine technology in collaboration with the Institute of Nanostructure Technology and Analytics (INA)[3] at the University of Kassel[4] as a scientific resource and B.Braun Melsungen AG[5] as a well established industrial partner in medicine technology. In this configuration several government grants haven been won and bilateral agreements settled at a total volume of 1.8 Million EUR. Up to now six university graduates have been employed and five Master's thesises completed on the topic of the so-called 'NanoNose'.

Figure 1.1 visualizes the organizational structure of the collaboration on the laser sensor system.

Current Status of Development

First experiments on the sensing principle have been carried out 2005 by H. Krause, who undertook efforts to force commercially available Distributed Feedback (DFB) Lasers into two mode oscillation. The oscillation state obtained turned out to be quite unstable and the output rather noisy due to the fact that the DFB Lasers used were poorly AR coated, which caused quite pronounced parasitic cavity effects. Also, these lasers consisted of a solid crystal and offered no possibility of introducing a sample into the cavity.

The experiments on bulk DFB Lasers were followed by the development of a first external cavity laser system, which featured a fiber coupler that distributed the incident light to two narrow band Fiber Bragg Gratings (FBG) in parallel alignment. Stable two mode oscillation

---

[1] www.gino-innovativ.de
[2] www.exist.de
[3] www.ina-kassel.de
[4] www.uni-kassel.de
[5] www.bbraun.com

1. Introduction

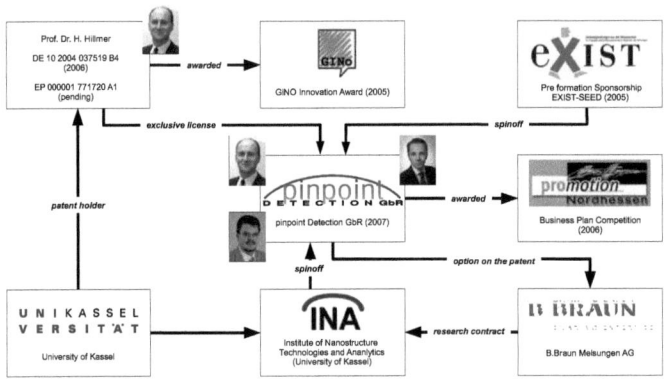

Figure 1.1.: Organigram of the partners involved in the registration and exploitation of patent DE 10 2004 037 519 B4.

was obtained. However, the setup still suffered from bad signal to noise ratio, as the fiber coupler caused loss of 75% on each round trip. The experiments performed on this system enabled the more recent work documented in [48], where the parallel FBG are replaced by a sequential setup, reducing cavity loss significantly. Further investigation on the behaviour of this laser under different operating conditions lay the foundation for the development of this work.

1.4. Objective and Chapter Outline

The focus of this work lies on the further development of the fiber based laser system described in [48] from a mere two mode laser to an actual sensing device. This includes the improvement of laser characteristics and the enhancement of the whole system in terms of functionality, to comply with the requirements of a sensing application. The final goal of this work is the proof of concept (POC) of patent DE 10 2004 037 519 B4 valid by first absorption experiments on Propofol and thereby reaching a big and important milestone on the way to an end user product that bases on the new and innovative sensing principle described in the patent.

1.4. Objective and Chapter Outline

The structure of this thesis is as follows:

- **Chapter 1** is this chapter and serves as an introduction to the scientific challenge of this project and presents an overview of competing fluid sensor principles.

- **Chapter 2** unrolls the fundamentals that the reader needs to understand in order to follow the chain of arguments and the significance of the results obtained in this work. It begins with an introduction to lasers in general and familiarizes the reader with the specific laser system at hand, explaining important parameters in laser characterization along the way. Finally, the concept of ICAS is presented in the context of the goal of this work, spectroscopic investigations on liquids.

- **Chapter 3** presents the laser system developed for the ICAS application described in patent DE 10 2004 037 519 B4. It is discussed under the perspective of suitability for the POC and the improvements and enhancements necessary for a successful POC on liquids are proposed. Components additional to the system in [48] are presented and characterized for compatibility with the application intended. It concludes with the presentation of the final laser system setup used for POC.

- **Chapter 4** quantifies the performance of the enhanced laser system and compares it to the previous one in the light of a sensing application. Its optical output is presented and its response to the variation of all freely adjustable parameters is investigated in detail. Additionally, the properties of the target substance, Propofol, are introduced to the reader and different solvents discussed in terms of compatibility with Propofol. Chapter 4 concludes with a suggestion of operating parameters for the POC measurements.

- **Chapter 5** is the scientific core of the work at hand. It presents the successful POC measurements on Propofol in solution of DiChloroMethane (DCM) and Acetone and their detailed evaluation. The laser systems suitability for ICAS is demonstrated and from the measurements obtained its sensitivity limit estimated.

- **Chapter 6** sums up the findings of this work, opening new questions that provide a guideline for future work on the topic of the two mode external cavity laser system developed in this work.

2. Fundamentals leading to an ICAS Experiment

This section familiarizes the reader with the fundamentals which are needed to understand this work. To provide a context to the upcoming chain of argumentation fig. 2.1 shows a very much simplified schematic setup of laser system complying to the patent DE 10 2004 037 519 B4.

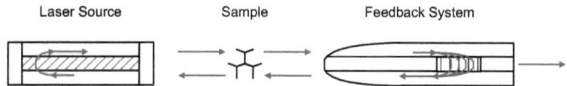

Figure 2.1.: Extremely simplified ICAS laser [38].

The schematics of the laser system depicted in fig. 2.1 consists of three logical components:

- A laser source, needed to provide laser radiation
- A sample to be measured
- A feedback system capable of wavelength selection

As the focus of this work is a proof of concept that requires a lot of operational flexibility, the utilization of commercial components is favored. Hence, the laser source used in this work is realized as a commercial semiconductor component and the rest of the system consists of very versatile fiber optics. As a result, the goal of the following introduction of fundamentals about lasing is unfolding the path to semiconductor lasers with external feedback. Where reasonable, the link to the big picture is established.

2. Fundamentals leading to an ICAS Experiment

2.1. The General Principle of Lasing

The following presentation of fundamentals begins with an outline of the lasing principle and introduces the different logical components needed to implement a laser, beginning with the very basic idea of how active material works and the possible transitions taking place in it. Subsequently, stimulated emission as a fundamental process in lasing is explained and the influence of a resonator on laser output discussed. It concludes with an introduction of the parameters used in the characterization of the laser system developed in this work.

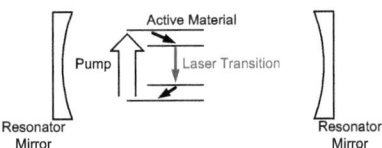

Figure 2.2.: Working principle of a laser, consisting of pumped active material between two mirrors: the pump energy applied to the active material establishes inversion, which provides amplification by stimulated emission of radiation.

Semiconductor lasers exhibit several properties which make them very desirable for spectroscopic applications. These include narrow band emission, high intensity, high coherence and a highly parallel beam. These properties of the light emitted are obtained by the generation of output light by stimulated emission. The three major functional parts of a laser are shown schematically in fig. 2.2:

- An energy source, the 'pump'
- Active material to provide amplification of laser radiation
- A resonator to provide optical feedback of the amplified laser radiation generated

2.1.1. Transitions in the Active Semiconductor Material

The active material of a laser is the component that provides the amplification of radiation. It requires at least three energy levels, the transition between two of which must be radiant. Figure 2.3 a shows such a basic three level system. The four level system, which is relevant to semiconductor lasers, is shown alongside in fig. 2.3 b.

2.1. The General Principle of Lasing

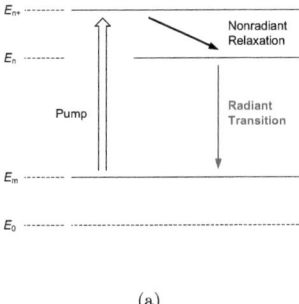
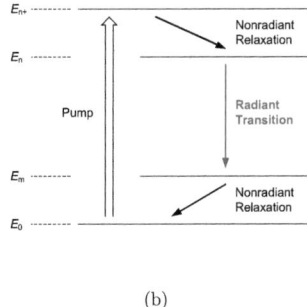

(a)   (b)

Figure 2.3.: Energy schemes of active material
(a) three level system, valid for ruby lasers [58]
(b) a four level system, valid for solid state lasers and semiconductor lasers.

There are three possible ways such multi level systems can interact with photons. They can absorb photons, emit them spontaneously or through a the process of stimulated emission.

In absorption, a photon's energy $E_{nm} = E_n - E_m$ is absorbed by a carrier in the energy level E_m, adding its energy to the energy of the carrier. This causes the carrier to raise from E_m to the excited state E_n.

After a finite time, which is the life time τ_n of the excited state, carriers in E_n spontaneously decay to the state E_m, emitting a photon that carries the energy difference E_{nm} between these states. Its direction and phase have no relation to the direction and phase of the initial photon. This process is called spontaneous emission.

In stimulated emission, which is the most important process for lasing, an incident photon of E_{nm} interacts with a carrier in the excited state E_n. Upon incidence it triggers its radiant decay from level E_n to E_m. The photon emitted during this process differs significantly from a photon emitted spontaneously. In particular, its direction, phase, wavelength and polarization are identical to the incident photon, making both photons indistinguishable.

Apart from transitions involving photons, also non radiant transitions can occur between energy levels, especially in solid state lasers, like semiconductor lasers. The energy released by such a transition is dissipated by other means, such as phonons.

11

2. Fundamentals leading to an ICAS Experiment

2.1.2. Inversion as the prerequisite for Lasing

For lasing to occur a domination of stimulated emission over all other transitions is required. This situation is ensured by continuously supplying sufficient energy to the active material to excite or 'pump' carriers from E_m to E_{n+}, depleting E_m of carriers. E_{n+} is an instable energy level with a short life time compared to E_n. Hence, carriers in E_{n+} quickly undergo a non-radiant transition from E_{n+} to E_n populating the metastable energy state E_n, where they reside. This situation, in which there are more excited states occupied than fundamental ones, is contrary to the naturally occurring distribution of carriers from thermal energy [47] and is called the inverted state, or inversion. It is a prerequisite for the amplification of radiation in the active material that is needed for lasing. Inversion is mainly dependent on the power of the pump, the life times of the excited states E_{n+} and E_n and the number of photons in the active material.

Incident photons carrying the energy E_{nm} will now most likely trigger stimulated emission from E_n. The source of these incident photons can be either an external source or a previous process of spontaneous emission in the active material. The emission spectrum created by such spontaneously emitted photons triggering stimulated emission is called Amplified Spontaneous Emission (ASE) and is in most cases undesired and regarded as a contribution to optical noise.

The magnitude of the amplification of spontaneously emitted photons by stimulated emission in the active material is called material gain G_{act}. The condition $G_{\text{act}} = 1$ defines the pump power at which the active material becomes transparent, and is hence called transparency excitation. In a semiconductor laser, material gain is reduced by geometry effects. This is accounted for my the effective gain G_{eff}, given by eq. 2.1 [39]. The loss from imperfect overlap of laser mode with waveguide in eq. 2.1 is expressed by the confinement factor Γ.

$$G_{\text{eff}} = \Gamma G_{\text{act}}, \qquad (2.1)$$

2.1.3. Mathematical Description of the Lasing Process

Mathematically the process described above are modeled based on transition probabilities, which can be expressed by the Einstein Coefficients $\mathscr{A}_{mn}$, $\mathscr{B}_{mn}$ and $\mathscr{B}_{nm}$. These transition probabilities are mostly dependent on the population N of the respective energy levels and the number of photons Φ in the laser cavity. The higher an energy level is populated, the more likely a transition out of that energy level becomes. Hence, the probability for absorption is

2.1. The General Principle of Lasing

given by the following expression [93].

$$\dot{N}_{m,\text{Abs.}} \sim -\mathscr{B}_{mn}N_m \tag{2.2}$$

For spontaneous emission the formula changes to

$$\dot{N}_{n,\text{Sp.}} \sim -\mathscr{A}_{mn}N_n. \tag{2.3}$$

The probability for stimulated emission is given by

$$\dot{N}_{n,\text{St.}} \sim -\mathscr{B}_{nm}N_n \tag{2.4}$$

and the probability for non-radiant decay by

$$\dot{N}_{n,\text{nr}} \sim -\mathscr{A}_{n+n}N_n, \tag{2.5}$$

where $\mathscr{A}_{n+n}$ is the Einstein Coefficient for spontaneous Emission from E_{n+} to E_n.

Under the assumption that the non-radiant relaxation transitions from E_{n+} to E_n and E_m to E_0 are fast enough to be neglected, the static and dynamic behavior of a four level laser can be derived from eq. 2.2-2.5. It is expressed by two rate equations for the number of inverted carriers $N = N_n - N_m$ and the number of photons in the cavity. It is given in eq. 2.6.

$$\begin{aligned}\dot{\Phi} &= \left(\mathscr{B}_{nm}L_{bb}N - {}^1\!/_{\tau_c}\right)\Phi \\ \dot{N} &= P - \mathscr{B}_{nm}\Phi N - \mathscr{A}_{nm}\cdot N,\end{aligned} \tag{2.6}$$

where L_{bb} is broad band cavity loss, P the pump rate and τ_c is photon life time.

Increasing stimulated emission in terms of eq. 2.2-2.5 means to increase $\dot{N}_{n,\text{St.}}$ by overpopulating energy level E_n relative to E_m. Then eq. 2.4 becomes larger than eq. 2.3 and the probability for stimulated emission increases.

2. Fundamentals leading to an ICAS Experiment

2.1.4. The Effect of the Resonator on Laser Emission

The domination of stimulated emission in the active material alone does not suffice to provide decent beam quality or high intensity of the laser radiation emitted. In order to obtain this the amplified radiation must pass the active material multiple times, so that an avalanche effect can occur by multiple re-amplification on each round trip. The avalanche effect is achieved by encapsulating the active material between mirrors or incorporating it into a ring in which the radiation travels. Any optical setup providing optical feedback in such a way is called resonator. Many different forms of resonators are known, i.e. Fabry-Pérot (FP) resonators, ring resonators based on total reflection inside a disc or fiber [80, 29, 46], resonators based on Distributed Bragg Reflectors (DBR) [54], just to name a few. The most simple case often applied in semiconductor lasers, the FP resonator shown in fig. 2.4 a, is used to illustrate the resonator's function in a laser, following along the lines of [36].

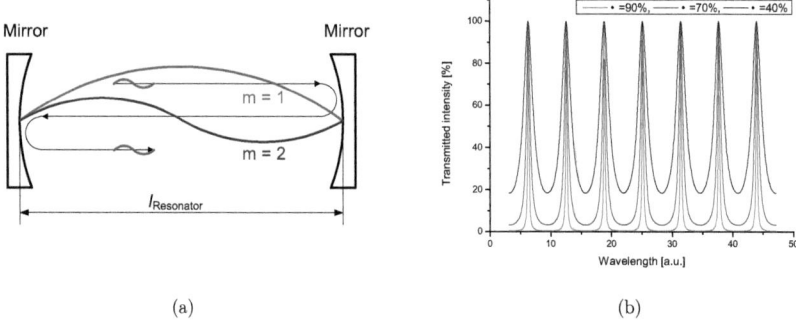

(a) (b)

Figure 2.4.: Wavelengths supported by the FP resonator
(a) Visualization of constructive interference of the fundamental and first order mode in the FP resonator
(b) Normalized output of a FP Resonator calculated by the transfer function given in eq. 2.8 for different mirror reflectivities.

The FP resonator consists of two parallel mirrors of equal reflectivity R which are placed at a defined distance in relation to each other. In order to be supported by the resonator, the radiation propagating inside must experience constructive interference with its reflection and develop a standing wave. This FP condition, given by eq. 2.7, is only fulfilled for discrete wavelengths. It can be translated into a calculated transmission spectrum of the FP resonator shown in eq. 2.8, plotted for different mirror reflectivities in fig. 2.4 b.

2.1. The General Principle of Lasing

$$n_{\text{eff}} l_{\text{Resonator}} = \frac{\lambda}{2m}, \tag{2.7}$$

where n_{eff} is the refractive index in between the mirrors, $l_{\text{Resonator}}$ is resonator length, λ wavelength and m the order of resonator mode, an integer.

$$\mathscr{I} = \frac{\mathscr{I}_0}{1 + {}^{4R}\!/\!{(1-R)^2} \cdot \sin^2{(\omega/2)}}, \tag{2.8}$$

where $\mathscr{I}$ is the transmitted intensity, $\mathscr{I}_0$ the incident intensity, R the reflectivity of the resonator mirrors and ω angular frequency of the radiation. The distance of the peaks in the spectrum shown in fig. 2.4 b is called the free spectral range FSR of the resonator, given by eq. 2.9

$$\text{FSR} = \frac{\lambda^2}{2l_{\text{Resonator}}} \tag{2.9}$$

The sharpness of an individual peak is accounted for by the FWHM. It is defined as the width of a given bell shaped function, in this case a Gaussian, at 50% of its maximum intensity, like illustrated in fig. 2.5

In logarithmic scales 50% corresponds to -3 dB. Hence, the FWHM can be found at 3 dB below the maximum of a given laser line. A small FWHM is most important for any application involving wavelength selection, such as DWDM or spectroscopy.

Together, FSR and FWHM deliver a measure for the optical quality of the FP resonator, the Finesse $\mathscr{F}$. It is defined as their ratio eq. 2.10 and depends solely on mirror reflectivity.

$$\mathscr{F} = \frac{\text{FSR}}{\text{FWHM}_{\text{Mode}}} = \frac{\pi\sqrt{R}}{1-R} \tag{2.10}$$

A different parameter often used to quantify the quality of laser resonators is the Quality factor (Q-factor), defined in eq. 2.11.

$$Q = \frac{\nu}{\Delta\nu} = \frac{\lambda^2}{\lambda \cdot \text{FWHM}} \overset{\text{FWHM} \ll \lambda}{\approx} \frac{\lambda}{\Delta\lambda}, \tag{2.11}$$

where ν is the frequency of the incident radiation.

It is a general measure for the ratio of the energy stored in an oscillator and the energy dissipated

2. Fundamentals leading to an ICAS Experiment

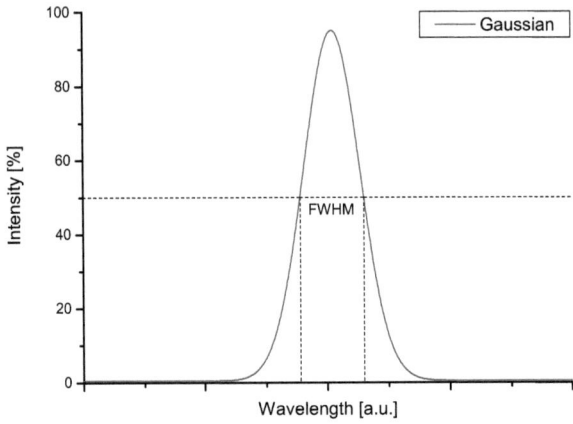

Figure 2.5.: Illustration of the FWHM of a normalized Gaussian function.

by it per cycle and accounts for its capability of conserving energy. The expression above assumes ideally symmetric peaks and otherwise undistorted optical output. In reality however, no resonator is perfect and subject to loss and distortion from imperfect mirrors [18], diffraction, interference, absorption or scattering and coupling, which each contribute to deviations from ideal optical output. This results in a decrease of Q-factor and Finesse as well as an increase in FWHM compared to the ideal case.

2.1.5. The Threshold Condition

Stable lasing occurs, when the light amplified by stimulated emission suffices to compensate for all optical loss occurring in the system of active material and resonator on each round trip. This occurs first for wavelengths that meet two criteria:

1. They have to be inside the wavelength region which receives sufficient gain from the active material.

2. They have to be a wavelength supported by the resonator

2.1. The General Principle of Lasing

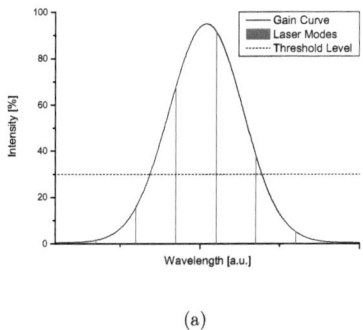
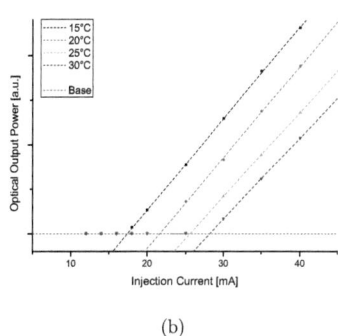

(a)   (b)

Figure 2.6.: (a) Threshold condition for laser modes.
(b) Measurement of the output intensity of a commercial laser diode as a function of injection current for different temperatures. The kink indicates threshold current.

Figure 2.6 a illustrates this condition schematically. Each one of the red laser modes calculated by the transfer function of the FP Resonator receives amplification determined by the gain of the active material. Those, which receive sufficient gain to exceed the threshold level, begin to lase. Laser threshold is defined as the pump power at which the gain from the amplification by stimulated emission overcomes total loss L_{total} in the resonator. Threshold gain G_{thr} is given by eq. 2.12 [28].

$$\Gamma G_{\text{thr}} = L_{\text{total}}, \qquad (2.12)$$

If this condition is satisfied for one or more modes, these modes consume any further supplied pump power and directly translate it into optical output.

Figure 2.6 b shows laser output as a function of pump power for a commercial laser diode[1] at different temperatures, recorded with a broad band power meter[2]. At pump powers below the threshold level, a small linear increase of output intensity with input power is observed, as inversion in the laser builds up. In fig. 2.6 b this curve is denoted as 'Base'. When lasing threshold is reached for one or more modes, the increase of output power remains linear, but the slope becomes much steeper than before, as all energy supplied to the laser pump is now consumed by the lasing modes and translated to optical output. The output power at which

---
[1] NEC NDL5801P-16
[2] Melles Griot Universal Power Meter

2. Fundamentals leading to an ICAS Experiment

these two linear functions intersect is taken as a measure for threshold current. It is one of the most important parameters in laser characterization.

It must be noted, however, that the gain curve is not stationary, as assumed in the considerations basing on fig. 2.6 a. Due to effects from temperature and carrier density in the active material it shifts in wavelength as described in sec. 2.2.1. This means that resonator modes which are located at gain maximum at low pump power are not necessarily the ones to reach lasing first.

2.1.6. Parameters Frequently Used in Laser Characterization

Laser threshold and lasing wavelength alone are not sufficient to characterize the output of a given laser comprehensively. Also the intensity ratio between the lasing mode and noise, as well as the width of the lasing mode has to be accounted for. These two characteristics are quantified by the Signal to Noise Ratio (SNR) and the FWHM, explained in sec. 2.1.4 of the laser emission.

Signal to Noise Ratio

SNR is defined as the ratio between the intensity of the lasing wavelength $\mathscr{I}_\text{Mode}$ and the intensity of the underlying optical noise $\mathscr{I}_\text{Noise}$. It depends mostly on the quality of the laser resonator. It is generally defined as

$$\text{SNR} = -\frac{\mathscr{I}_\text{Mode}}{\mathscr{I}_\text{Noise}}. \tag{2.13}$$

As the output spectra of lasers are most commonly plotted on logarithmic scales due to the large span in magnitude, eq. 2.13 changes to

$$\text{SNR}_\text{log} = \mathscr{I}_\text{Mode} - \mathscr{I}_\text{Noise} \tag{2.14}$$

A theoretical investigation of the dependence of SNR on coupling efficiency, laser crystal reflectivity, and FBG reflectivity of an external cavity semiconductor laser, similar to the one in sec. 2.2.3, shows that SNR mainly depends on the coupling coefficient between semiconductor crystal and fiber [54]. For increased SNR it is favorable to increase coupling efficiency and FBG reflectivity in the system. Reflectivities from other sources, such as the laser crystal, decrease SNR.

2.2. Resonator Designs Interesting for ICAS

A typical value for SNR obtained from a state of the art telecommunication laser is 45 dB [33].

2.2. Resonator Designs Interesting for ICAS

To give an understanding of the conceptual origin of the semiconductor laser based laser system employed in this work, the following section gives an overview of three laser resonators, which each contribute to the setup used in sec. 3.1. Firstly, a FP semiconductor laser is introduced. Secondly, a modification that makes it a DBR laser is presented. Finally, the concept of external cavities is explained, which leads to the idea for the external cavity implemented in the laser system developed in this work.

2.2.1. The Fabry-Pérot Resonator

A simple semiconductor laser resonator is the FP resonator. It consists of a p-/i-/n-doped (PIN) diode as active material, a resonator consisting of two reflecting facettes of the semiconductor crystal and an electrical current source for pumping, as shown in fig. 2.7 a.

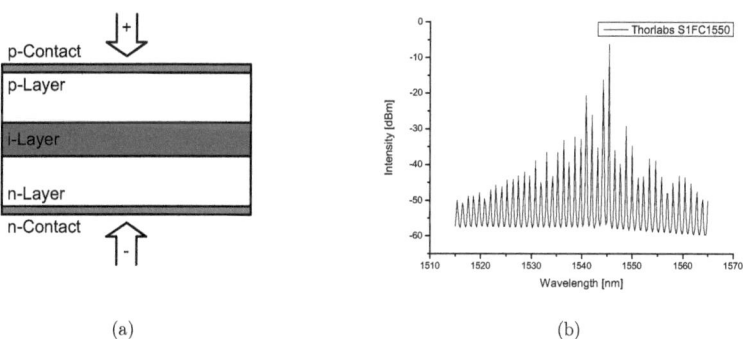

Figure 2.7.: (a) Schematic cross section of a FP semiconductor laser diode with electrical contacts.
(b) Example of the optical output spectrum of Thorlabs S1FC1550 FP laser source, provided by the vendor.

The quality of the resonator in the case of such a simple FP semiconductor laser is rather poor, as its mirrors only consist of the mere facettes of the semiconductor crystal. Reflection is based

2. Fundamentals leading to an ICAS Experiment

on the difference in refractive index between semiconductor material and surrounding air; this means that the reflectivity of the facettes of such a simple device is given by Fresnel reflection of the semiconductor to air interface, which is noted in eq. 2.15 [11].

$$R_{\text{Fresnel}} = \left[\frac{(n_{\text{h}} - n_{\text{l}})}{(n_{\text{h}} + n_{\text{l}})}\right]^2, \qquad (2.15)$$

where n_{h} is the refractive index of the semiconductor crystal and n_{l} the lower refractive index of the surrounding air. Assuming $n_{\text{h}} = 3.5$ and $n_{\text{l}} = 1$, the reflectivity of the semiconductor air interface returned by eq. 2.15 is as small as 30% [22]. Finesse as a measure of resonator quality for this type of laser is 2.45 according to eq. 2.10 and eq. 2.15.

Besides the difference in the resonator mirror, FP semiconductor lasers also differ from the most simple case of a three level laser, shown in fig. 2.3 a. Their energy level $E_{\text{n+}}$ is one of many inside the conduction band of the semiconductor and an additional level E_0 depicted in fig. 2.3 b, which lies inside the valence band. Its purpose is to deplete E_{m} by means of non radiant transitions as quickly as possible to avoid re-absorption of photons previously emitted in the lasing process.

The FP semiconductor laser bases on the PIN junction. Its band diagram in the simple case of the homo junction laser diode under forward bias is shown in fig. 2.8 a, where an intrinsic semiconductor layer is sandwiched between a p-doped layer on the left and an n-doped layer on the right [22]. As a result of bias, electrons are injected into the i-layer from the right and holes are injected from the left. In the i-layer they recombine under stimulated emission of radiation. The PIN homo junction is electrically leaky and lacks a waveguide to confine photons emitted to the i-layer, where most of the recombination of carriers takes place. Electrical leakage is solved by cooling the device. Optical loss is unaffected by that. The problem is ultimately solved utilizing a combination of high band gap material with low refractive index for the p- and n-doped layers and low band gap material with high refractive index for the i-layer to confine carriers and photons to the i-region. This structure is called the double hetero structure PIN junction and was firstly proposed by [3] and [50]. Its band diagram under forward bias is shown in fig. 2.8 b [39].

For simplicity, its working principle is described for electrons and the n-layer, but in fact a description for holes and the p-layer is just as valid. Applying forward bias to the structure shown in fig. 2.8 b drives the electrons through the conduction band of the n-layer into the i-layer. At the n/i-interface they encounter a small potential barrier, which results from technological limitations of doping and bias. This is small enough to be tunneled through and poses no

2.2. Resonator Designs Interesting for ICAS

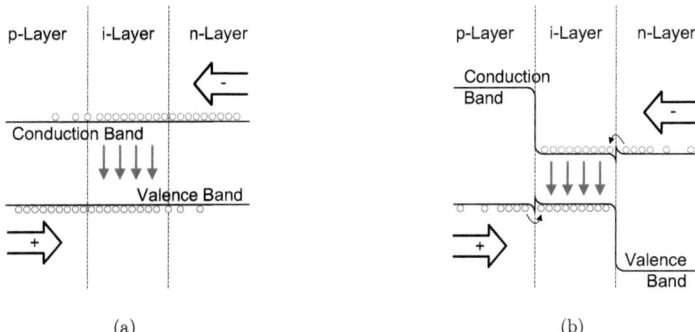

Figure 2.8.: (a) Band diagram of a PIN homo junction [22],
(b) Band diagram of the forward biased double hetero structure PIN junction [39].

obstacle to the electrons. They reach the i-layer to which they are confined as a result of the major potential barrier at the i/p-interface. The resulting buildup of electrons in the conduction band and holes in the valence band of the i-layer provides the inversion needed for stimulated emission to occur.

Due to electrical and optical confinement it allows cw operation of a laser diode at room temperature and significantly lower threshold current than a laser diode with a homo junction. Initially this emission is more or less broad band, depending on the specific band gap of the semiconductor material applied. The radiation traverses inside the resonator, where those wavelengths corresponding to the maxima of the resonators transfer function, eq. 2.8, can survive best and trigger further stimulated emission in the i-layer. Finally, the emission condenses to very few narrow laser lines. A typical output spectrum of such a laser is shown in fig. 2.7 b.

Two particular specialties apply in semiconductor lasers, that influence optical output properties. These are

1. The dependence of the gain curve on the temperature of the semiconductor crystal (red shift) and

2. The dependence of the gain curve on carrier density (blue shift).

2. Fundamentals leading to an ICAS Experiment

Red Shift

When the temperature of the semiconductor crystal is increased, the maximum of the gain curve of the active material is shifted towards longer wavelengths. This thermal red shift is caused by different effects. One the one hand, the thermal expansion of the crystal elongates the resonator and thereby also moves the position of the modes supported by it. The magnitude of the shift is mainly dependent on the thermal expansion coefficient of the semiconductor material applied. On the other hand, the electrostatic lattice potential is reduced by the increase of mean atomic distance [26], which results in a reduction of the band gap energy E_{Gap} and a corresponding longer wavelength. An empirical formula is derived in [98], given in eq. 2.16.

$$E_{\text{Gap}}(\Theta) = E_{\text{Gap}}(0) - \frac{f_1 \Theta^2}{\Theta + f_2}, \qquad (2.16)$$

where E_{Gap} is the gap energy at 0K, Θ is temperature and f_1 and f_2 are material specific fitting parameters. The reduced smaller band gap also increases the refractive index of the material, resulting in another contribution to a shift to longer wavelengths. For the Indium Gallium Arsenide (InGaAs) material system this is investigated in [43]. An example of the combination of the effects is measured for a commercial semiconductor laser diode[3] and shown in fig. 2.9 a.

Blue Shift

As a counterpart to thermal red shift, the blue shift from increasing carrier density is observed in semiconductor material. The more carriers are present in the energy bands of the semiconductor, the more carriers occupy states that are further away from the band edge. This Fermi level band filling enables transitions not only from the band edge, but also from inside the bands. Hence, the transitions span a larger energy difference and the wavelength of the radiation emitted is shifted to shorter wavelengths. Additionally, higher carrier density causes a bigger interaction between the carriers, influencing their effective mass. This, however, is a complicated many particle problem, whose detailed explanation lies outside the scope of this work. Details on the two effects mentioned above is found in [56].

What remains is that below threshold an increase of pump current results in an increase of carrier density, which coincides with a blue shift of the gain curve. When threshold is reached, all extra pumping energy is translated to optical output. No further blue shift is observed above threshold, as the carrier density remains constant then. Further increase of pump energy raises

---
[3] NEC NDL5801P-16

2.2. Resonator Designs Interesting for ICAS

the temperature of the device resulting in a red shift of the gain curve. This effect is mostly suppressed as a result of dissipating the heat responsible by means of heat sinks. Measurements of the blue shift of a commercial semiconductor laser diode[4] at 20°C is shown in fig. 2.9 b.

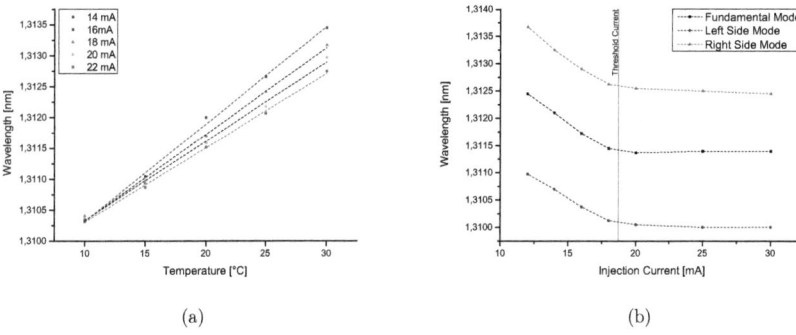

Figure 2.9.: (a) Measurement of the red shift of the output wavelength of a commercial semiconductor laser under increasing temperature.
(b) Measurement of the blue shift of the output wavelength of a commercial semiconductor laser under increasing injection current.

2.2.2. Exploiting Bragg Reflection for Laser Resonators

A significant enhancement of the very simple laser design of the FP Laser shown in in sec. 2.2.1 is the edge emitting DBR laser. Its main improvement consists of the application of DBR mirrors to the facettes of the semiconductor crystal. Their working principle bases on the transformation of Bragg reflection to optics. They can be tailored to inhibit very high wavelength dependent reflectivity and low absorption.

Figure 2.10 a shows a schematic setup of such an edge emitting DBR laser. The active material corresponds to the FP laser depicted in fig. 2.8 a, sec. 2.2.1. The mirrors, which are shown to the left and right of the active material are the DBR mirrors. They consist of a multilayer system of two alternating materials, with difference in refractive index and optical thickness of $\lambda/4$. Radiation penetrating such a DBR under normal incidence experiences partial reflection at each layer interface due to the refractive index contrast. Wavelengths matching the period of

---
[4]NEC NDL5801P-16

2. Fundamentals leading to an ICAS Experiment

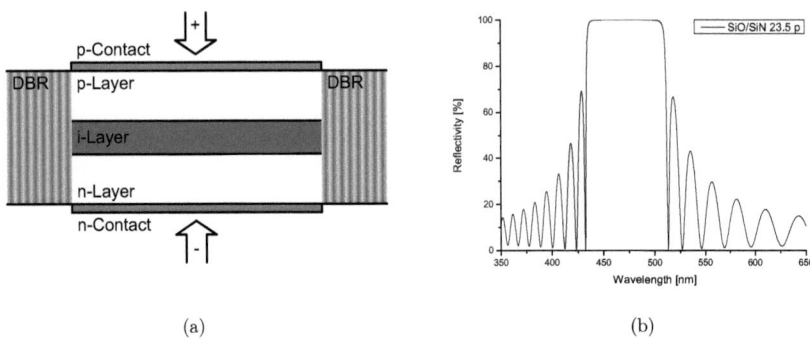

Figure 2.10.: (a) Schematic cross section of a FP semiconductor laser diode with DBR mirrors. (b) Calculated spectral reflectivity of a DBR consisting of 23.5 periods of $\lambda/4$-layers SiO_2/SiN.

the DBR experience constructive interference in the direction of reflection; reflectivity of over 99.99% is easily achieved in this way.

Mathematically the reflection spectrum is modeled best by the Transfer Matrix Model, which accounts for the transfer function of each individual layer and the resulting reflection spectra of the complete multilayer system. A detailed derivation is presented in [73]; an example is shown in fig. 2.10 b. The main feature of the spectrum is its high flat center section. Its width is referred to as the mirror's stop band and can be tailored to meet any individual requirements depending on the availability of materials and the technological limitations of layer deposition. It has been shown in [57] that under neglect of absorption the reflectivity R at design wavelength of such a multilayer system deposited on a solid substrate can be derived to

$$R_{\mathrm{DBR}} = \left(\frac{1 - (n_{\mathrm{h}}/n_{\mathrm{l}})^{2p} (n_{\mathrm{h}}^2/n_{\mathrm{eff}})}{1 + (n_{\mathrm{h}}/n_{\mathrm{l}})^{2p} (n_{\mathrm{h}}^2/n_{\mathrm{eff}})} \right)^2, \tag{2.17}$$

where $2p+1$ is the number of layers in the multilayer system, n_{h} and n_{l} are the high and low refractive indexes of the layers, respectively, and n_{eff} the effective refractive index of the active material of the laser diode in between the DBR. Obviously, R is dependent on the refractive index contrast $\Delta n = n_{\mathrm{h}} - n_{\mathrm{l}}$ and the number of periods of the DBR. Hence, reflectivity can be increased by either increasing Δn or, if high contrast materials are not suitable, by increasing p.

2.2. Resonator Designs Interesting for ICAS

The width FWHM$_{\text{DBR}}$ of the DBR stop band can be derived to the following expression, also following [57]:

$$\text{FWHM}_{\text{DBR}} = \frac{2}{\pi} \sin^{-1}\left(\frac{n_{\text{h}} - n_{\text{l}}}{n_{\text{h}} + n_{\text{l}}}\right), \qquad (2.18)$$

In contrast to its reflectivity R at design wavelength, FWHM$_{\text{DBR}}$ is independent of the number of periods - the higher the contrast, the wider the stop band. Considering absorption and yield the implementation of DBR with very few layers is favorable. In this case, high refractive index contrast is necessary to obtain high reflectivity from eq. 2.17. Compared to a FP laser, a DBR laser has a much higher Finesse solely due to its better resonator mirrors. Assuming a reflectivity of only 90% [44] eq. 2.10 returns a value of 29.8, which is ten times as high as the one obtained for a FP laser with cleaved facets only.

2.2.3. Tailoring Laser Emission with External Cavities

External cavity semiconductor lasers receive their feedback from a resonator that extends significantly outside the semiconductor crystal as such. A schematic setup of an external cavity semiconductor laser in Littrow configuration [30] is shown in fig. 2.11.

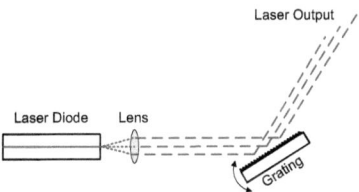

Figure 2.11.: Example of an external cavity semiconductor laser with a grating as the wavelength selective element (Littrow configuration) [30].

Just like in the case of the lasers mentioned in sec. 2.2.1 and 2.2.2 also in the external cavity semiconductor laser the diode itself serves as active material. However, at least one of its facettes is Anti Reflection (AR) coated to couple light into an external feedback system [28]. The cavity as such can be designed in many different ways. Common external cavities use collimation into free space optics [102, 32] or coupling into a fiber based feedback system [32]. The individual feedback system implemented depends largely on application. Especially in respect to tuning characteristics an external feedback system inhibits a few attractive properties

2. Fundamentals leading to an ICAS Experiment

compared to the lasers presented in sec. 2.2.1 and 2.2.2, which can only be tuned by means of temperature or injection current as discussed in sec. 2.2.1. Both methods are limited to a small tuning range of a few nm and cannot reliably eliminate mode hopping. For certain applications, such as optical spectroscopy, this is not tolerable [103]. External cavity semiconductor lasers allow a variety of different tuning methods to be applied. Any tuning mechanism, which can be implemented into the external cavity can be used, i.e. gratings [102], prisms [94, 79, 16], etalons [99] or various types of filters based on interference [106, 10], birefringence [4] or acousto optic modulators [62, 24]. This allows the exploitation of the full gain bandwidth for phase continuous wavelength tuning.

Due to their more complex setup compared to the FP or DBR Laser external cavity semiconductor lasers also suffer from a few problems, such as additional loss, undesired interference effects and increased sensitivity to mechanical influences. Additional loss originates mainly from mirror loss between the cleaved facet ends of the semiconductor crystal, loss caused inside the external cavity and loss due to the imperfect overlap between the optical field of the external cavity with the active region.

Other facettes besides those of the crystal also play an important role in such a system as they form undesired (parasitic) cavities, which can significantly change the output spectrum of a laser system. Careful attention has to be paid to that in the design of external cavities. Where possible, AR coating should be applied to the facettes inside the cavity in order to reduce their influence. This is discussed in detail in sec. 3.2.4.

Special care has to be taken in the mechanical layout of an external cavity semiconductor laser. The coupling of light between the crystal and the external feedback system requires high precision, as mechanical influences such as vibration on impact cause permanent misalignment, resulting in an increase of loss or the shifting of parasitic cavity modes. Especially in external feedback systems involving free space light propagation this can be critical.

2.3. Intra Cavity Absorption Spectroscopy in the NIR

This section deals with the interaction of Electro Magnetic (EM)-radiation and light. The nature of the rotational, vibrational and rotational-vibrational molecule absorption spectra of molecules is discussed and different line broadening mechanisms introduced following the lines of [100]. Subsequently, the origin of continuous absorption spectra of liquids are explained. Finally their detection by means of ICAS is discussed and the advantages and disadvantages of common ICAS setups evaluated.

2.3.1. EM-Active Transitions in NIR: Rotational-Vibrational Spectra

In classical electrodynamics light is regarded as propagating electric and magnetic vectors oscillating on perpendicular planes orthogonal to the direction of propagation. If an electrical charge is exposed to such EM-radiation, its electric component forces it into oscillation. In return, an oscillating charge, like a Hertz dipole, emits EM-radiation. This also applies to the interaction of EM-radiation with the electrical charges inside atoms or molecules, where the excitation of different transititions are distinguished by the frequency or wavelength range of the radiation they interact with. An overview is given in tab. 2.1.

Radiation	Wavelength m	Frequency Hz	Wavenumber cm^{-1}	Energy kJ/mol	Transition
Radio	$10^2 - 1$	$3 \cdot 10^6 - 3 \cdot 10^8$	$10^{-4} - 10^{-2}$	$10^{-6} - 10^{-4}$	Nucleus Spin
µ-Wave	$1 - 10^{-2}$	$3 \cdot 10^8 - 3 \cdot 10^{10}$	$10^{-2} - 1$	$10^{-4} - 10^{-2}$	e^- Spin
µ-Wave	$10^{-2} - 10^{-4}$	$3 \cdot 10^{10} - 3 \cdot 10^{12}$	$1 - 10^2$	$10^{-2} - 1$	Rotation
IR	$10^{-4} - 10^{-6}$	$3 \cdot 10^{12} - 3 \cdot 10^{14}$	$10^2 - 10^4$	$1 - 10^2$	Vibration
UV/Vis	$10^{-6} - 10^{-8}$	$3 \cdot 10^{14} - 3 \cdot 10^{16}$	$10^4 - 10^6$	$10^2 - 10^4$	Outer e^-
X-Ray	$10^{-8} - 10^{-10}$	$3 \cdot 10^{16} - 3 \cdot 10^{18}$	$10^6 - 10^8$	$10^4 - 10^6$	Inner e^-
γ-Ray	$10^{-10} - 10^{-12}$	$3 \cdot 10^{18} - 3 \cdot 10^{20}$	$10^8 - 10^{10}$	$10^6 - 10^8$	Nucleus States

Table 2.1.: Overview of the EM-active transitions observed in atoms and molecules with associated magnitude and experimental method [91].

The wavelength range of $10^{-4} - 10^{-6}$ is the range of the Near Infra Red (NIR), where the laser system developed in this work is operated. This EM-radiation can interact with vibrational and rotational molecule oscillations. In the following their origin and properties are explained for the case of cold low pressure gases, where discrete lines are observed that can be precisely explained theoretically.

2. Fundamentals leading to an ICAS Experiment

Rotational Spectra

Figure 2.12 shows a simplified one dimensional model of a rotating molecule with a permanent dipole moment μ. EM-radiation propagating in the z-direction can interact with this molecule by energy exchange. It is known from quantum mechanics that only discrete amounts of energy can be exchanged [23]. For the angular momentum of molecules this fact is accounted for by the quantum number J. If energy is absorbed by molecule, J is increased; if energy is released, J is decreased. In any case the molecule undergoes a transition from one rotational energy level E_{rot} into another E'_{rot}.

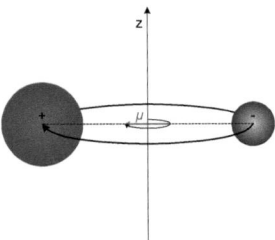

Figure 2.12.: The generation of an oscillating dipole moment by rotation of an electric dipole.

Rotational energy E_{rot} is discrete and given by the eigenvalues of the Schrödinger equation for a rigid rotor, eq. 2.19 [23]

$$E_{\text{rot}}(J) = B \cdot hcJ(J+1),$$
$$B = \frac{h}{8\pi^2 c\theta} \tag{2.19}$$

where h is Planck's constant, c the velocity of light and θ is the molecule's moment of inertia; B is called rotation constant. Transitions are only allowed for the condition $\Delta J = \pm 1$, leading to absorption lines that are equidistant, which can be expressed in wavenumber $\tilde{\nu}$ notation eq. 2.20, [100]

$$\Delta \tilde{\nu} = \lambda^{-1} = 2B(J+1) \tag{2.20}$$

In a real system this equidistance is not strictly observed as higher rotations result in greater

2.3. Intra Cavity Absorption Spectroscopy in the NIR

inter atomic distances from centrifugal force, which influences B. Hence, for larger B the distance of absorption peaks decreases. Also, rotation is not limited to the molecule as a whole; in larger molecules also individual groups can rotate relative to the rest of the molecule. In this case the pure rotational spectra derived here are distorted by the coupling of the inner rotation to the outer rotation. This is observed as a fine structure in molecule spectra [19].

The energy of pure rotational spectra is on the order of $2 \cdot 10^{-22}$ Joule[5], which corresponds to $\tilde{\nu}_0 = 20.8\,(J+1)$ cm^{-1} and lies in the range of $\lambda = 4.76 \cdot 10^{-4}$ m.

Vibrational Spectra

Besides rotation, molecules can also carry energy in inter atomic vibration. For a simple molecule consisting of only three atoms the three vibrational modes possible are shown in fig. 2.13a, b and c. As the existence of a time dependent change of dipole moment is required for interaction with EM-radiation, the symmetric stretching mode shown on the left is EM inactive. The asymmetric stretch and the bending mode are active.

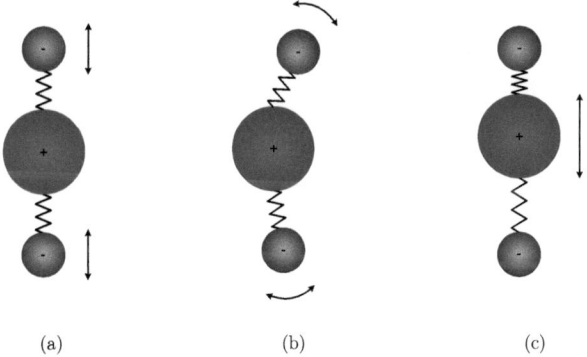

(a) (b) (c)

Figure 2.13.: Vibration modes of a molecule consisting of three atoms
(a) In the symmetric case the vibration does not cause a change in dipole moment and no interaction of EM-radiation is possible.
(b), (c) Asymmetric modes cause a change in dipole moment and enable interaction with EM-radiation.

Vibrational spectra are based on the solution of the Schrödinger equation for the Morse potential

---
[5]for HCl

2. Fundamentals leading to an ICAS Experiment

shown in fig. 2.14, a potential based on the harmonic oscillator, but enhanced by a distortion that accounts anharmonicity. Its eigenvalues for a diatomic molecule are given by eq. 2.21.

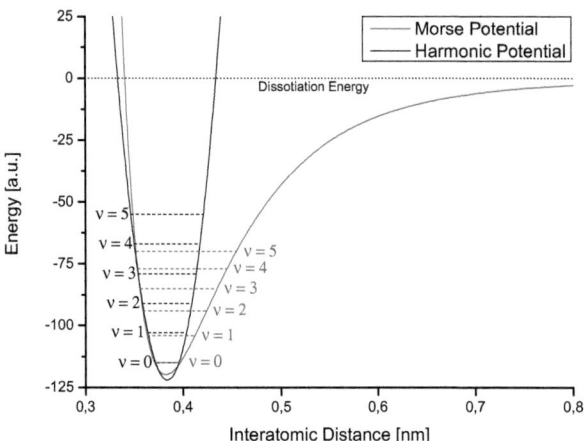

Figure 2.14.: Eigenvalues of the Morse potential in comparison to those of the harmonic potential. Spacing is reduced for the asymmetric Morse potential and converges to 0 at the dissociation energy of the molecule.

$$E_{\text{Morse}}(j) = E_{\text{harm.}}(\nu) - hc\tilde{\nu}_0 \Omega \left(j + 1/2\right)^2,$$
$$E_{\text{harm.}}(j) = hc\tilde{\nu}_0 \left(j + 1/2\right)$$
(2.21)

where j is order of vibration, and ν_0 is the oscillation frequency of the molecule, given by eq. 2.22 and Ω is a factor introduced to account for the anharmonicity of the Morse potential.

$$\nu_0 = \frac{1}{2\pi}\sqrt{\frac{\kappa}{m_{\text{red.}}}},$$
(2.22)

where κ is Hook's constant of the atomic bonding and $m_{\text{red.}}$ reduced mass. Transitions are only allowed for the condition $\Delta j = \pm 1, \pm 2, \pm 3, ...$, which leads to an expression for the energy difference of adjacent vibrational lines, given in eq. 2.23

2.3. Intra Cavity Absorption Spectroscopy in the NIR

$$\Delta E = \tilde{\nu}_0 \left[1 - 2\Omega \left(j + 1\right)\right] \quad (2.23)$$

ΔE decreases for increasing j. From this decline in energy difference between adjacent vibrational modes the dissociation energy E_{Diss} of a molecule can be calculated. It is reached when molecule vibrations become too intense for the molecule binding to keep the molecule together. It is given by eq. 2.24

$$E_{\text{Diss}} = \frac{\tilde{\nu}_0}{4\Omega}. \quad (2.24)$$

The energy of vibrational excitation is about two orders of magnitudes larger than the energy of pure rotational excitation. Its wavenumber is on the order of $\tilde{\nu}_0 = 2700\,(j+1)$ cm^{-1}[6] and lies in the range of 10^{-6} m [100].

Rotational-Vibrational Spectra

Experimentally vibrational spectra usually turn out not to be infinitely sharp as each vibrational line is broadened by the series of its corresponding rotations. The resulting spectra are called rotational-vibrational spectra and have the morphology of the exemplary spectrum shown in fig. 2.15.

Its energy can be approximated by addition of eq. 2.21 and eq. 2.19 and results in the discrete energy levels given by eq. 2.25

$$\Delta E = hc \cdot \tilde{\nu}_0 \left(j + {}^1\!/{}_2\right) + B \cdot J\left(J + 1\right) \quad (2.25)$$

Transitions between rotational-vibrational states are allowed only if they equally fulfill both transition conditions to J and j mentioned above. The combined condition for rotational-vibrational transitions, eq. 2.26, leads to a division of the rotational-vibrational spectra into two groups of lines, which are labeled P- and R-branch. As transitions for $\Delta J = 0$ are forbidden, the transition for $\tilde{\nu}_0$, the so called Q-branch, is not observed.

---

[6]for HCl

2. Fundamentals leading to an ICAS Experiment

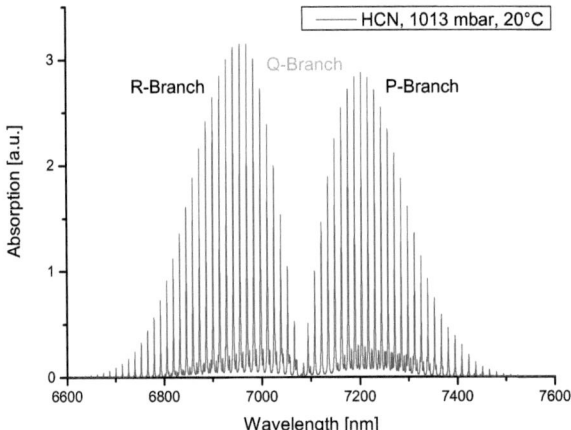

Figure 2.15.: Rotational-vibrational absorption spectrum of HCn, calculated from HITRAN database The P- and R-branch are clearly observed, the Q-branch is not observed because it corresponds to a forbidden transition.

2.3. Intra Cavity Absorption Spectroscopy in the NIR

$$\Delta j = +1, \Delta J = \pm 1$$
$$\Delta j = -1, \Delta J = \pm 1$$
(2.26)

2.3.2. Relevant Sources of Absorption Line Broadening

For the POC measurements in sec. 5 of this work, absorption is measured on liquids. Due to liquid phase and temperature, their absorption lines are not composed of discrete lines, as described for the ideal case of cold gases above. Different broadening mechanisms turn them into continua as a result of the vastly increased interaction of the molecules in the liquid. The ones relevant for NIR spectroscopy on liquids are explained in the following.

Natural Linewidth

The natural line width is the fundamental line width of any emission- or absorption line originated in a transition between two different energy levels of finite life time. In classical physics such a transition can be characterized by a damped wave, given in eq. 2.27 [26]

$$x(t) = x_0 \cdot \exp^{-(\gamma/2)t} \cdot \left[\cos(\omega t) + \frac{\gamma}{2\omega} \sin(\omega t) \right],$$
(2.27)

where x denotes oscillation amplitude, ω and ω_0 are angular frequency and angular eigenfrequency of the oscillator, respectively, and γ the damping factor of the oscillation. Frequency characteristics are obtained by Fourier transformation $\mathrm{FT}(x(t))$, and its normalized intensity $\mathscr{I}(\nu)$ is then given by eq. 2.28

$$\mathscr{I}(\omega) = |F(x(t))|^2 = \eta_{\mathrm{nat}} \cdot \mathscr{I}_{\mathrm{nat}}(\omega),$$
$$\mathscr{I}_{\mathrm{nat}}(\omega) = \frac{\gamma/2\pi}{(\omega - \omega_0)^2 + (\gamma/2\pi)^2}$$
(2.28)

where $\mathscr{I}_{\mathrm{nat}}$ is the Lorentzian envelope of the function and η_{nat} a factor introduced for normalization.

The life time of the transition is given by the time at which the corresponding oscillation decreases to $1/\gamma$ of its initial amplitude. It is called the natural life time $\tau_{\mathrm{nat.}}$ of the excited

2. Fundamentals leading to an ICAS Experiment

state, which for spontaneous emission is dominated by the associated Einstein coefficient $\mathscr{A}_{nm}$ introduced in sec. 2.1. This is expressed in eq. 2.29

$$\tau_{\text{nat.}} = \gamma^{-1} = A_{nm}^{-1}, \tag{2.29}$$

Together with Heisenberg's uncertainty principle $\Delta E \cdot \Delta \tau \geq \hbar$ this means that every emission caused by an energy transition of finite life time must have a finite spectral line width, if the life time of the connected excited state is also finite. Its frequency characteristics are given by a Lorentzian envelope, whose FWHM can be calculated from eq. 2.28 and is given by eq. 2.30

$$\Delta \nu_{\text{nat}} = (2\pi \tau_{\text{nat.}})^{-1} = \frac{\mathscr{A}_{nm}}{2\pi} \tag{2.30}$$

Collision Broadening

In its origin and morphology, pressure broadening is very similar to natural line width. It is also a life time induced line broadening mechanism and sometimes also called collision broadening, as the origin of the finite life time of states is found in the collision of atoms and molecules.

To adapt the expression for natural line broadening given in eq. 2.30 to collision broadening the relevant life time $\tau_{\text{nat.}}$ is replaced with the collisional life time $\tau_{\text{col.}}$ [5]. It is given by eq. 2.31.

$$\tau_{\text{col.}} = \zeta^{-1} \tag{2.31}$$

The quantity ζ is the collision frequency, defined as

$$\zeta = \sigma v_{\text{Molecules}} \mathscr{N}, \tag{2.32}$$

where σ represents the collision cross section, $v_{\text{Molecules}}$ the relative mean speed of the molecules involved in the collision processes, both given in eq. 2.33 and $\mathscr{N}$ the number of molecules per unit volume. σ and $\bar{c}_{\text{rel}}$ are defined in eq. 2.33.

$$\sigma = \pi d^2$$
$$\bar{c}_{\text{rel}} = \sqrt{\frac{8 k_B \Theta}{\pi m_{\text{red.}}}}, \tag{2.33}$$

2.3. Intra Cavity Absorption Spectroscopy in the NIR

where d represents collision diameter and ranges in the order of magnitude of molecule diameters. $k_B\Theta$ is the Boltzmann factor and $m_{\text{red.}}$ the reduced mass of the collision pair. Subsequent substitution of eq. 2.30 by eq. 2.31, eq. 2.32 and eq. 2.33 leads to the final expression for the line broadening $\Delta\nu_{\text{col}}$ by collision processes

$$\Delta\nu_{\text{col}} = \sqrt{\frac{2k_B\Theta}{\pi m_{\text{red.}}}} \cdot \mathcal{N} d^2 \qquad (2.34)$$

As for all life time based effects, collision broadening also results in a line shape very similar to a pure Lorentzian and is given by eq. 2.35 [26].

$$\mathscr{I}_{\text{col}}(\omega) = c_{\text{col}} \cdot \frac{\left[1/2\left(\gamma_n + \gamma_{in}\right) + \mathcal{N}\bar{c}_{\text{rel}}\varsigma_b\right]^2}{\left[\omega - \omega_0 + \mathcal{N}\bar{c}_{\text{rel}}\varsigma_s\right]^2 + \left[1/2\left(\gamma_n + \gamma_{in}\right) + \mathcal{N}\bar{c}_{\text{rel}}\varsigma_b\right]^2}, \qquad (2.35)$$

where γ_n and γ_{in} are damping constants of the amplitude oscillations by irradiation of emission and inelastic collisions, respectively, and ς_b and ς_s measures for line width broadening and line shift due to elastic collisions, respectively.

In addition to pure line broadening from elastic and inelastic collisions, elastic collisions also shift the position of spectral lines as a result of phase changes during the collision process. Equation 2.35 accounts for both effects.

Doppler Broadening

Doppler broadening is caused by the thermal distribution of molecule velocities in an absorber or emitter and is independent on any life time effects as the broadening mechanisms previously described. The transition frequency of a moving source is shifted by its relative velocity v in respect to the stationary receiver according to eq. 2.36

$$\nu' = \nu\left(1 + v/c\right). \qquad (2.36)$$

From Maxwell-Boltzmann statistics the number of atoms per velocity interval can be calculated and the intensity distribution of the line in the frequency domain derived [26]. It is a Gaussian function given by eq. 2.37

$$\mathscr{I}_{\text{Doppler}}(\omega) = \eta_{\text{Doppler}} \cdot \exp^{-\left(c(\omega-\omega_0)/\omega_0 \cdot \sqrt{\mu/2k_B\Theta}\right)^2}, \qquad (2.37)$$

2. Fundamentals leading to an ICAS Experiment

where ω equals angular frequency and ν_0 is angular eigenfrequency. The FWHM of eq. 2.37 delivers the line broadening by Doppler effect in eq. 2.38

$$\Delta\nu_{\text{Doppler}} = \frac{\nu_0}{c}\sqrt{\frac{8 \cdot \ln(2) \cdot k_B \Theta}{m_{\text{red}}}}. \tag{2.38}$$

For a detailed account of the nature of the rotational, vibrational and rotational-vibrational spectra of molecules please refer to [5], [41] and [100].

Liquid Phase Spectra

The question remaining is how precisely the individual impact of the broadening mechanisms mentioned above is in turning the gas phase spectra derived in sec. 2.3.1 into continua. For this purpose a volume V of liquid is compared with a volume of gas. For comparison eq. 2.30, 2.34 and 2.38 are re-listed in simplified form in eq. 2.39; all constant factors are substituted into arbitrary constants $\mathscr{E}$, leaving the independent variables remaining.

$$\begin{aligned}\Delta\nu_{\text{nat}} &= \mathscr{E}_{\text{nat}} \\ \Delta\nu_{\text{col}} &= \mathscr{E}_{\text{col}} \cdot \sqrt{\Theta} \cdot \mathcal{N} \\ \Delta\nu_{\text{Doppler}} &= \mathscr{E}_{\text{Doppler}} \cdot \sqrt{\Theta}\end{aligned} \tag{2.39}$$

Assuming an adiabatic transition from gas to liquid phase it is obvious that natural line width $\Delta\nu_{\text{nat}}$ remains unaffected. $\Delta\nu_{\text{col}}$ and $\Delta\nu_{\text{Doppler}}$ both increase with increasing temperature, which equally shows in collision and Doppler broadening. The change of density, however, additionally leads to an increase of line width from collision broadening. As a result it can be stated that collision broadening is the effect dominating the elimination of discrete spectral lines visible in gas phase when transiting to liquid phase. Especially the rotational states vanish completely. Phenomenologically this can be understood when regarding the adiabatic transition as a reduction of the mean free path length under maintained impulse of particles. Collisions occur significantly more often and lead to a much more homogeneous distribution of energy and the elimination of any sharp energy states in the ensemble. The absorption spectra of the substances important for the POC of patent DE 10 2004 037 519 B4 attempted in this work are presented in sec. 4.2.

2.3. Intra Cavity Absorption Spectroscopy in the NIR

2.3.3. Intra Cavity Absorption

A straightforward method for measuring absorption characteristics of material is to expose it to EM-radiation of intensity $\mathscr{I}_0$ and measure the intensity reduction in a single pass by the transmitted intensity $\mathscr{I}$. Lambert-Beer law, given in normalized form by eq. 2.40 then allows to calculate the absorption coefficient.

$$A = 1 - T = 1 - \exp^{-\alpha_{\text{eff}} \cdot l_{\text{Absorber}}}, \qquad (2.40)$$

where A represents absorption, T transmission, α_{eff} the absorption coefficient and l_{Absorber} absorber length.

For wavelength resolved absorption measurements commonly a combination of a broad band light source and a subsequent monochromator is applied. However, due to their spectral properties also semiconductor lasers are attractive alternatives for this purpose. Besides narrow line width, good tunability and the option of tailoring gain characteristics to almost any spectral region they offer the possibility of increasing sensitivity significantly. This happens as a direct consequence of the nature of the lasing process itself, if the absorber is incorporated directly into the laser resonator. The method is called ICAS.

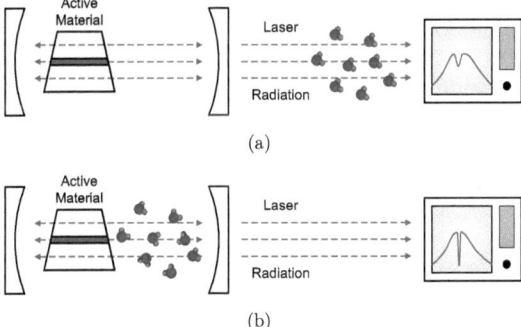

Figure 2.16.: (a) Schematic setup of a conventional absorption experiment: a laser is used as light source and its emission subjected to absorption outside of the resonator after amplification.
(b) Schematic setup of an ICAS experiment: high sensitivity is achieved by placing the absorber inside the resonator.

Figure 2.16 a shows a conventional spectroscopic setup compared to an ICAS setup shown in

2. Fundamentals leading to an ICAS Experiment

fig. 2.16 b. In the ICAS setup the radiation generated by the active material of the laser system propagates back and forth between the two resonator mirrors. Modes fulfilling the resonance condition (eq. 2.7) of the resonator experience constructive interference and receive gain. A wavelength selective absorber placed inside the resonator influences net gain according to its absorption characteristics. The relative intensity of the resulting modes reflects the absorption features in the absorber. ICAS is much more sensitive than conventional single pass spectroscopy. A measure for sensitivity is given by the effective path length l_{eff}, defined by eq. 2.41 [8], below.

$$l_{\text{eff}} = \beta \frac{A}{\alpha_{\text{eff}}} \qquad (2.41)$$

$$\beta = l_{\text{Absorber}}/l_{\text{Resonator}}$$

where α_{eff} is the absorption coefficient of the material. β denotes the filling factor of the absorber, a measure for the fraction of the laser cavity filled by absorbing material, where l_{Absorber} is absorber length and $l_{\text{Resonator}}$ resonator length. l_{eff} can also be expressed in the form of eq. 2.42, an approach often used for theoretical considerations in literature.

$$l_{\text{eff}} = c t_{\text{sat}}, \qquad (2.42)$$

where c is the speed of light and t_{sat} the spectral saturation time of the absorption signal. In this work, no time resolved measurements are taken.

The enhancement of sensitivity in ICAS over the conventional single pass method results from two main effects:

- **The resonator effect**, which increases sensitivity by repeated amplification of the absorption signal in multiple passes of the active material. This effect is independent of mode interaction.
- **Mode competition** has by far the stronger effect on sensitivity. It is largest for strictly homogeneously broadened gain and is discussed in the following sec. 2.3.4.

Inhomogeneous contributions to gain broadening or other effects in favor of mode coexistence like spontaneous emission, spatially inhomogeneous intensity distribution of standing wave type modes or effects caused by spatial and temporal inhomogenities in the active material limit mode competition and thereby also the sensitivity of an ICAS system. This is discussed

2.3. Intra Cavity Absorption Spectroscopy in the NIR

in more detail in the following sec. 2.3.4.

2.3.4. Sensitivity Limitations of ICAS

A prerequisite for mode competition is strong mode coupling by i.e. homogeneously broadened gain. That means that laser gain originates equally from all atoms of the active material in the same way. In this case all resonator modes lying within the gain bandwidth of the active material initially grow under increasing pumping power. When they reach threshold, the mode receiving most gain increases in intensity over the others as a result of the avalanche character of laser amplification, depleting the common inversion in the active material. That results in the quenching of all other modes and the subsequent condensation of the laser spectrum to the single strongest mode.

In strictly inhomogenously broadened gain the total gain is composed of several independent entities in the active material providing gain, i.e. all atoms of the same relative velocity from thermal energy. In this case only the modes that receive gain from the same entity engage in mode competition. Modes that receive gain from different entities of atoms in the gain media cannot interact in terms of mode competition. They coexist, which limits sensitivity of ICAS [20].

In [8] the theoretical limit of sensitivity for ICAS with a four level multi mode laser is derived in dependence on a parameter X_q that accounts for various distortions from the strictly homogeneous case. The calculation bases on an extension of the rate equations in eq. 2.6, given for the mean values of the photon numbers Φ_q in mode q and the number of inverted carriers state N shared by all modes in eq. 2.43 [9].

$$\begin{aligned}\dot{\Phi}_q &= -L_{bb}\Phi_q + \mathscr{B}_{nm,q} N\Phi_q - \alpha_q c\Phi_q + X_q \\ \dot{N}_n &= P - \mathscr{A}_{nm}N - N\sum_q \mathscr{B}_{nm,q}\,\Phi_q,\end{aligned} \quad (2.43)$$

where L_{bb} is broad band cavity loss, $\mathscr{A}_{nm}$ the Einstein Coefficient for spontaneous emission, $\mathscr{B}_{nm,q}$ Einstein Coefficient for stimulated emission for mode q, α_q the absorption coefficient for mode q and P the pump rate.

The theoretical limit of sensitivity l_{eff} is obtained by substitution of the absorption signal A in eq. 2.41 by the stationary absorption signal A_q^s derived from eq. 2.43. A_q^s is given in eq. 2.44

2. Fundamentals leading to an ICAS Experiment

$$A_q^s \cong \frac{\alpha_q c \Phi_q^s}{X_q}, \tag{2.44}$$

where Φ_q^s is the stationary photon number after laser spectrum stabilization. The limit of sensitivity l_{eff} is given by eq. 2.45

$$l_{\text{eff}} = \frac{c \Phi_q^s}{X_q}. \tag{2.45}$$

The parameter X_q is determined by the specific type of distortion of the laser regarded. For the omnipresent distortion of mode coupling by spontaneous emission $X_q = B_q N^s$, where N^s is the population of the excited state after laser spectrum stabilization.

$$l_{\text{eff}} = \frac{c \Phi_q^s}{\mathscr{B}_{\text{nm},q} N^s} \tag{2.46}$$

The effective path length, given in eq. 2.46 from spontaneous emission, given in eq. 2.46, increases with increasing photon number and pump power. Experimental evidence of this for diode lasers is found in [7], [31].

Distortion of mode coupling predominantly by Rayleigh scattering, i.e. in the optical fiber of the laser system investigated in this work, is accounted for by substituting $X_q = S_{\text{Rayleigh}} M_q^s$, where S_{Rayleigh} denotes Rayleigh scattering. The resulting l_{eff}, eq. 2.47, is independent of photon number or pump rate, which in the general case do not correlate directly, if line width changes under varied pump power due to saturation effects. They are hence discussed here separately. How that fact reflects in l_{eff} is found in [8].

$$l_{\text{eff}} = \frac{c}{S_{\text{Rayleigh}}} \tag{2.47}$$

Distortions of laser emission by nonlinear effects, such as four wave mixing or stimulated Brillouin scattering are accounted for by $X_q = D \cdot \left[M_q^s\right]^2$, where D is an independent function. An approximation for l_{eff} dominated by these nonlinear contributions is given by eq. 2.48.

$$l_{\text{eff}} = \frac{c}{D \Phi_q^s} \tag{2.48}$$

In general, smaller X_q means stronger coupling between individual modes of the laser system.

2.3. Intra Cavity Absorption Spectroscopy in the NIR

As a result, increased sensitivity is to be expected from an ICAS experiment for smaller X_q. However, X_q is not the only indicator for the strength of mode interaction. Other factors, such as mode spacing [42] or number of modes [52, 34], also strongly influence mode interaction significantly.

Although multi mode operation is the most favorable approach in terms of sensitivity, ICAS also works with single mode lasers [77]. The sensitivity of a single mode ICAS setup is still much larger than in conventional single pass measurements due to the resonator effect. However, it is significantly smaller than in the multi mode case, as there is of course no mode competition. For increased sensitivity from mode competition effects, there must be more than one mode [96]. An increase of laser modes however, does not necessarily mean an increase to large numbers. The sensitivity of ICAS with two modes is theoretically investigated in [52]. It is argued that there is no significant difference of ICAS sensitivity to be expected between the two mode and an extreme multi mode situation. From the point of view of laser system design this fact creates a big benefit, as discussed in detail in the following sec. 2.3.5.

2.3.5. The Advantages of Two Mode ICAS

This section introduces the configuration of a typical experimental setup used for ICAS and compares the practical consequences of multi mode ICAS with single mode ICAS. Further, it suggests two mode ICAS as a means of circumventing the problems inherit to the two other cases. The focus of the evaluation lies on the implementation and application of such systems for the purpose described in patent DE 10 2004 037 519 B4: the possibility of miniaturization and the potential for mass market.

Multi Mode ICAS

Figure 2.17 a shows the schematics of an ICAS system in multi mode configuration. The broad band gain provides amplification for all modes supported by the resonator. The absorber inside the cavity suppresses those modes which coincide with its absorption features. This can be sets of discrete lines or broader features as in liquids. Absorption and gain together determine the net gain that each one of the resonator modes experiences. It is the parameter that determines their output intensity. A comparison of the initial output spectra of the laser without absorber and the output spectra obtained with the absorber in the resonator delivers the spectral absorption characteristics of the absorber.

2. Fundamentals leading to an ICAS Experiment

Regarded from the point of view of precision and sensitivity multi mode operation is very attractive. Due to mode competition the multi mode configuration is extremely sensitive [34] and, within the limitations of laser gain bandwidth, has the great advantage of providing the full absorption spectrum of an absorber in one single measurement. Very attractive lasers for this reason are dye lasers [34, 84] or fiber lasers [16], as they provide very flat gain over a large wavelength range.

However, the high fidelity offered by such a system comes at a price, and that is cost and delicacy. A setup as depicted in fig. 2.17 a involves several very expensive components. First of all flat broad band gain requires either a dye laser system or fiber laser system pumped by a large array of tailored high intensity pump diodes. Both systems have low efficiency and are quite costly. The dye laser system is additionally very delicate in setup and maintenance. To keep the gain flat and free of fluctuation, it has to be monitored and adjusted continuously.

To evaluate the optical output a multi mode system requires parallel evaluation of all mode intensities in the full bandwidth of laser gain. This parallel evaluation of broad band optical spectra necessarily involves an Optical Spectrum Analyzer (OSA), another rather expensive and delicate system in itself. Apart from the cost of individual components the multi mode setup in fig. 2.17 a is quite complex. To deliver reproducible and reliable results it must be precisely calibrated to provide flat output characteristics that are stable over time. The gain material must be designed to maintain stable amplification characteristics, which requires in situ monitoring of absorption free output spectra. The mechanical stability of a multi mode resonator also has to be taken care of, as changes in the resonator immediately reflect in laser output.

For the introduction of an ICAS based sensor into the mass market miniaturization is a key factor, as smaller devices have much greater possibilities of implementation. For the system shown in fig. 2.17 a miniaturization is highly limited by complexity of the setup and the requirements of the active material. Dye lasers require a dye beam, which is very difficult to be implemented in a miniaturized form. Fiber lasers offer slightly better options in this respect but the array of pumping diodes and the fiber path needed for amplification pose limitations.

Hence, multi mode ICAS systems are rather limited to applications where size, complexity or cost are not critical. In such fields, however, highest precision can be expected.

2.3. Intra Cavity Absorption Spectroscopy in the NIR

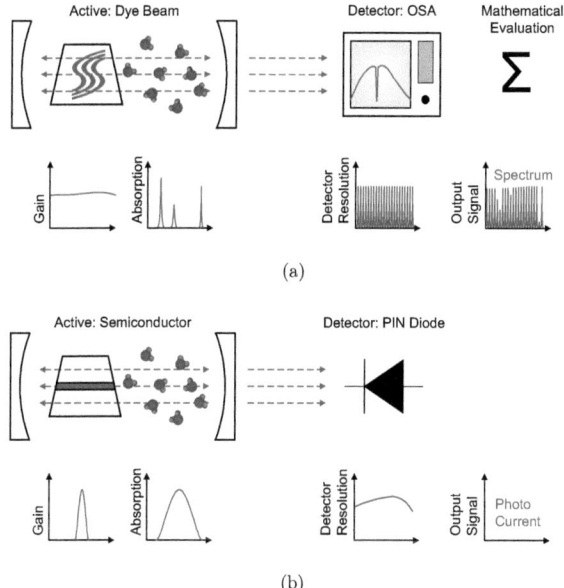

Figure 2.17.: Multi mode ICAS vs. single mode ICAS
(a) In the highly sensitive multi mode case complex and expensive dye or fiber lasers and high resolution measurement equipment is required.
(b) In the single mode case much more cost efficient devices like diode lasers and photo detectors can be applied, but sensitivity is significantly lower.

2. Fundamentals leading to an ICAS Experiment

Single Mode ICAS

In single mode ICAS only one mode is used for the detection of the absorption features of an absorber in the resonator. To obtain information on absorption at more than just one discrete wavelength a tunable laser source must be used. Single mode ICAS is generally less sensitive than multi mode ICAS due to the lack of mode competition. However, single mode ICAS systems, as shown schematically in fig. 2.17 b, are a lot less complex than the multi mode ICAS systems described above.

The reduced complexity mainly results from the limited gain bandwidth these systems require. To obtain lasing on only one single mode also narrow band gain is sufficient. Hence, no in-situ calibration of gain flatness is needed. Single mode output also greatly reduces the complexity of the resonator implemented. Where multi mode systems need to be designed to provide equal reflectivity for all wavelengths of the gain bandwidth, a resonator designed for one single tunable mode only needs flat reflectivity within the small tuning range of this one mode. The greatest simplification compared to multi mode systems, however, is a direct result of the simplicity of the output of a single mode laser system; optical data evaluation is completely obsolete, as the OSA in fig. 2.17 b can be replaced by a photo detector. Its electrical signal is proportionate to the total power it is irradiated with. In a single mode system this power is in any case allocated to the one known mode of the laser system, which qualifies the electrical signal of such a broad band photo detector a sufficient means of laser output quantification.

The much simpler setup compared to multi mode systems greatly eases miniaturization by integration in semiconductor material. Tunable semiconductor laser diodes and photo detectors are available in a vast wavelength range, as semiconductor band gap can be tailored by material composition [39]. Integrated waveguides are fabricated by definition of refractive index from doping and resonator mirrors can be implemented as DBR and tailored for the spectral reflectivity characteristics desired. The limiting factor for miniaturization is the sample chamber, which has to be designed in regard to the absorption characteristics of the absorber targeted. The cost of such an integrated device is much smaller than extensive macroscopic solutions like the multi mode ICAS setup described above and therefore much more suitable for the mass market.

However, the great reduction of complexity also leaves its marks in the performance of a device. In the case of the single mode ICAS setup this means reduction of sensitivity as a result of the lack of mode competition and the limitation to a rather narrow wavelength range. In order to increase bandwidth of a sensor based on single mode ICAS the fabrication of arrays is a necessity.

Two Mode ICAS

A possible solution to combine the advantages of single- and multi mode ICAS is utilization of ICAS under mode competition between two modes, proposed in DE 10 2004 037 519 B4. Here, two modes are used in the particular configuration that one mode is affected by a discrete absorption line from a wavelength selective absorber in the laser cavity, from which the other stays unaffected. As a result from resonator effect and mode competition, the mode experiencing absorption is significantly suppressed compared to the other. In theory this approach is predicted to have similar fundamental sensitivity as multi mode ICAS with a large number of laser modes [52]. Due to the reduction of mode numbers to two, however, data evaluation is significantly simpler.

In contrast to the optical evaluation of a large multi mode spectrum as required in the general case of multi mode ICAS, two mode ICAS can utilize a simple photo detector just as for single mode ICAS. Although a photo detector is not capable of wavelength selective measurements, the refined method of frequency resolved evaluation of photo current can be exploited to retrieve information on the intensity distribution between two defined modes. The method analyzes the noise of the photo current in the frequency domain, which is very characteristic for the single- or two mode state. The frequency spectra obtained in this manner are called relative intensity noise and directly correlate with the relative optical intensity of the two modes of the ICAS system, as demonstrated in [49]. Relative intensity noise as an indicator for the intensity difference between the two modes of the system can hence serve as sensor signal, making the expensive optical evaluation of laser output obsolete.

Generally, detailed relative intensity noise measurements are similarly difficult to obtain from a photo detector as optical spectra are to obtain from a laser source. The method requires an electronic spectrum analyzer, as the optical evaluation of laser output requires an OSA. However, the features in relative intensity noise that enable conclusions about the modal oscillating state of a laser system are so coarse that they can potentially be evaluated by a tailored integrated circuit. Details on the evaluation of this matter are found in [82].

As a result, a two mode system is claimed to combine the advantages of multi mode and single mode ICAS systems. It is potentially as sensitive as fully multi mode ICAS systems and offers a much less complex setup that is significantly easier to integrate into a miniaturized device. Hence, it is a promising approach to bringing ICAS to a wide range of commercial applications

2. Fundamentals leading to an ICAS Experiment

outside of laboratories. A preliminary macroscopic laser setup that enables stable two mode oscillation is developed in [48]. In the following section it is presented as the basis of this work.

3. Laser Setup Evolution

3.1. Where this work begins: The Status Quo

3.1.1. Laser System Presentation

The starting point of this work is the laser system, developed and characterized in [48]. Regarded in the light of a sensing application, however, several enhancements turn out to be necessary. In this chapter the laser system as it existed at the beginning of this work is introduced and the enhancements implemented described.

Figure 3.1 [89] shows the initial laser setup upon which this work bases. It is an external cavity DBR resonator based on the one introduced in sec. 2.2.3. Its specialty is that it facilitates two cavities in one fiber system and is capable of stable oscillation on two wavelengths at the same time.

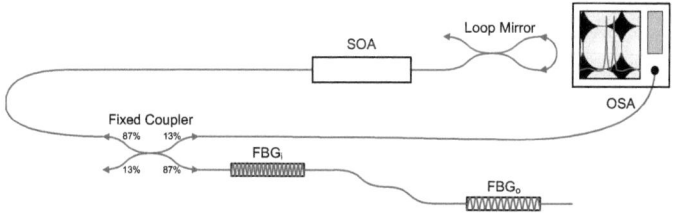

Figure 3.1.: Starting point of this work: the laser system from [48].

It consists of a Semiconductor Optical Amplifier (SOA)[1], one 87:13 fixed ratio fiber coupler[2], two FBG[3], and a loop mirror[4]. The SOA serves as gain medium, and two FBG as wavelength

---
[1] InPhenix IPSAD1501-L213, Specification in Appendix
[2] AOS GmbH custom made
[3] AOS GmbH custom made
[4] AOS GmbH custom made

3. Laser Setup Evolution

selective elements. The SOA enables stable multi-wavelength amplification and thereby serves as an optical source.

By applying a current to the SOA broad band radiation is emitted, amplified and coupled into the Single Mode Fiber (SMF) system attached. 50% of the radiation coming from the SOA is directly guided to the loop mirror on the right and fully returned. Resultingly, the loop mirror can be regarded as a broad band resonator mirror, with very high reflectivity at any wavelength.

Light exiting the SOA into the fiber system to the left first passes through a fixed ratio fiber coupler, that transmits 87% of its intensity. The remaining 13% are coupled out of the system to measurement equipment. FBG_i and FBG_o to the left of the fiber coupler in fig. 3.1 serve as resonator mirror for each of the two laser modes M_i and M_o, respectively. FBG_i provides very narrow band reflection on wavelength λ_i and is transparent for light of all other wavelengths, i.e. λ_o, which passes FBG_i unhindered and is reflected by FBG_o. Light of wavelength that corresponds neither to λ_i or λ_o passes both FBG unhindered and is coupled out of the system.

Hence, there are two cavities implemented in this system: cavity C_i with its fundamental mode M_i lasing on wavelength λ_i is formed by the loop mirror and FBG_i. Cavity C_o with its fundamental mode M_o lasing on wavelength λ_o is formed by the loop mirror and FBG_o. Both share the same active material, implemented by the SOA.

A detailed description of this system and its components can be found in [48].

As a clear definition for a starting point of this work the existing system is characterized and its parameters listed below.

3.1.2. Analyzing System Emission for Suitability in ICAS

A typical output spectrum of the laser system is shown in fig. 3.2. The dominant features are two laser modes on a sloped baseline. Their center wavelengths are 1541.98 nm and 1542.47 nm, tunable in range of about 2 nm each and correspond to the center wavelengths of the FBG. M_o at 1542.47 nm is about 3 dB more intense than M_i, due to the fact that they do not receive the same amount of gain. As they are spectrally placed on the slope of the gain curve, M_o, which is closer to gain center wavelength, receives more gain and is thus more intense.

Both modes show shoulders on their left slope. The origin of these is not clear, but it is assumed that they might be the effect of parasitic cavities in the laser setup. Additional to the two main modes at 1541.98 nm and 1542.47 nm, two side modes are observed at 1541.53 nm

3.1. Where this work begins: The Status Quo

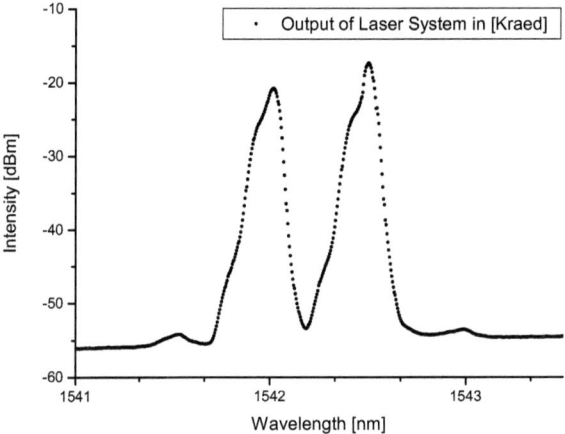

Figure 3.2.: Optical output measured for the laser system developed in [48]. M_i and M_o are found at 1541.98 nm and 1542.47 nm with FWHM of 0.08 ± 0.01 nm and (0.06 ± 0.01) nm, respectively. Mode intensity differs by 3.6 dB. Side modes are found at 1541.53 nm and 1543 nm.

3. Laser Setup Evolution

and 1542.99 nm. According to [48] they result from degenerate four wave mixing in the SOA, a nonlinear effect occurring in semiconductor lasers [64]. The intensity ratio between the center of M_i and M_o and these side modes from four wave mixing can be taken as measure for SNR, as defined in 2.1.5. It is roughly 35 dB in a typical operating scenario of this system. The higher the injection current of the SOA, the smaller SNR.

Line width is (0.08 ± 0.01) nm and (0.06 ± 0.01) nm for M_i and M_o, respectively. The threshold current I_{th} of this system is 91.5 mA for M_o and 92.35 mA for M_i and depends largely on the total loss caused by the fiber coupler. As it is passed twice per round trip, 24% intensity is lost.

3.1.3. Introduction of the Wavelength Tuning Mechanism

Wavelength tuning in the laser system described in [48] is implemented by tuning FBG reflectivity in the wavelength domain. The FBG themselves are glued to blocks of Aluminum, as shown in fig. 3.3 a. Heating these blocks results in thermal expansion and thereby in mechanical tension on the FBG. They are stretched and subjected to thermal expansion. As a result, their periodicity, which determines the wavelength reflected, is varied. This is described mathematically in [105] and has been simulated in [89].

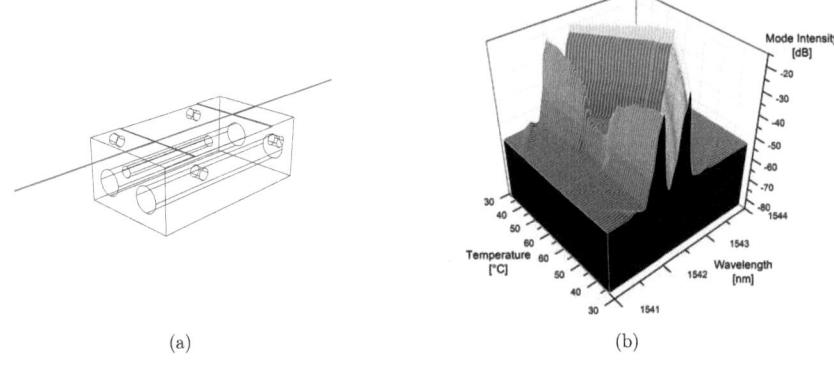

Figure 3.3.: FBG tuning mechanism
    (a) The FBG (red) is glued onto an Aluminum block (black). Heating the block causes a shift of its reflectivity to longer wavelengths due to thermal expansion.
    (b) Measured laser output under FBG tuning between 33°C and 65°C.

Figure 3.3b shows the output of the laser system under tuning of M_o at a rate of 0.034 nm/°C. Measurements have been taken after waiting for the output to stabilize. At maximum mode distance a breakdown in lasing is observed for M_i due to the increasing gain difference for the two modes at greater mode distance.

Mainly as a result of limited thermal conductivity between the heating elements and the fiber itself, and also the spatially inhomogeneous dissipation of thermal energy to the environment, the tuning of wavelength requires a finite settling time t_λ until a stable output wavelength is obtained. This time is documented in [89] and amounts to about 6 h, which seems long at first but poses no problem in the laboratory environment, as the system is typically stabilized over night. Stable output in this case means the state, where the drift in output wavelength from heating the Aluminum blocks becomes comparable to the fluctuations in output wavelength due to the inhomogeneous dissipation of thermal energy to the environment. An example of this situation in the initial laser system is shown in fig. 3.4.

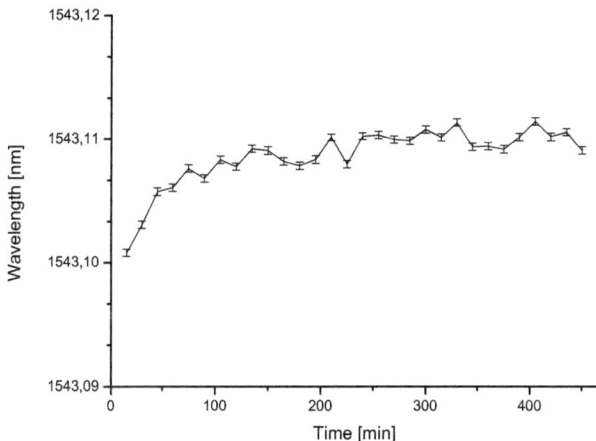

Figure 3.4.: Measured wavelength stability of the laser system documented in [89]. Fluctuation of 0.01 nm are observed after stabilization.

3. Laser Setup Evolution

3.2. Introducing the Modifcations to Obtain a Functional ICAS Setup

In order to make quantitative spectroscopy measurements on liquids with a laser system based on the one described in section 3.1.1, several enhancements are necessary for obtaining reliable results. This section deals with these enhancements, beginning with the implementation of variable fiber couplers to enable the equalization of mode intensities and adjusting threshold current. Subsequently, the introduction of a collimator arrangement to contain a sample chamber is described and optimizations to the thermal properties of the FBG feedback system presented. Finally the effect of improvements of the fiber connections and their effect on mode line width are discussed.

3.2.1. Equalizing Intensities and Tailoring Threshold

The sensing principle employed in this work bases mainly on the fact that absorption in the resonator disturbs the sensitive intensity equilibrium between two oscillating modes competing for gain. Initially a system has to be applied that allows an intensity equilibrium to be established at any wavelength position desired. In the system developed by [48] this is possible in case both modes are placed in the same distance of the gain maximum, where both receive the same amount of gain. However, for a sensing application that alone is not very useful, as the placement of the modes has to be chosen according to the substances to be detected, not according to the gain of the active material. One mode must be placed in a spectral position which receives low or no absorption by the target substance, the other must be placed in a spectral position that receives high absorption. Hence, the position of the modes is determined by the target substance. The resonator has to allow the establishment of an intensity equilibrium at this position without respect to the position of the gain curve of the active material.

For this purpose variable fiber optic couplers[5] are introduced, V_i as replacement for the existing fixed ratio coupler and a second one, V_o, which allows to equalize the oscillating modes no matter where on the gain curve they are placed. Additionally, variable couplers enable continuous wavelength tuning of the gain curve and adjustments to threshold current.

Figure 3.5a and b show the setup from [48] and the setup enhanced in this work, respectively. The difference that is regarded in this section is the presence of the two variable couplers V_i

---

[5]Fiberpro TC1410

3.2. Introducing the Modifcations to Obtain a Functional ICAS Setup

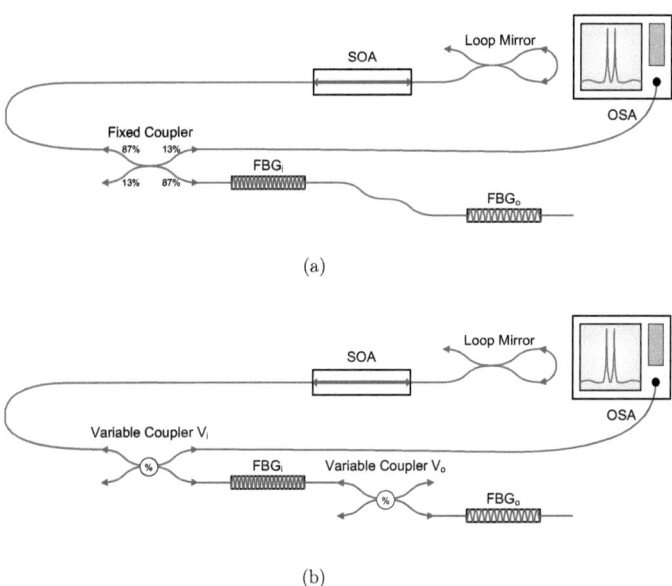

Figure 3.5.: (a) Laser system [48] with one fixed ratio coupler.
(b) Implementation of two variable couplers: V_i enables tailoring of threshold current and V_o equalization of mode intensities.

3. Laser Setup Evolution

and V_o in the improved setup. The fixed ratio coupler in fig. 3.5a is replaced by V_i and an additional coupler, V_o, is inserted between the two FBG. Both variable couplers have identical specifications and can be adjusted from $0 - 100\%$ coupling ratio by a knob with a scale of arbitrary units. They are used to introduce adjustable loss into the laser system. Note that coupler V_i affects both cavities C_i and C_o, whereas coupler V_o only affects cavity C_o.

By changing the coupling ratio of V_o the loss of C_o can be varied, while the loss in C_i remains unchanged. Regarding V_o as a logical unit with FBG_o, a variation in coupling ratio can be considered as a variation in the reflectivity of FBG_o. Or, from another point of view, V_o can be regarded as a logical unit with the wavelength dependent gain for this cavity, which means that a variation in coupling ratio can be considered a variation in the gain of M_o. Either way, an increase of coupling ratio decreases the intensity $\mathscr{I}_o$ of M_o relative to I_i of the M_i.

By configuring the system in such a way that M_o is the one closer to the gain maximum of the SOA, where it naturally receives higher gain than M_i, coupling ratio on V_o can always be used to re-equalize both modes. As a consequence, coupler V_o is used to force both modes into the, from a sensitivity point of view [52], much desired equilibrium.

The variable coupler V_i affects both cavities. Loss introduced by it will influence the intensities of M_i and M_o. Like above, regarding it as a logical unit with the FBG system, consisting of FBG_i, V_i and FBG_o, coupler V_i can be used to vary the total reflectivity on the side of the FBG in the laser system. This effect is similar to tunable broad band loss.

In section 2.1.5 the dependence of laser threshold on broad band loss, which by introduction of the variable coupler V_i is now tunable, is described. It can be exploited to adjust the threshold carrier density of the laser system. For the special case of an electrically pumped semiconductor laser this closely correlates with the position of the gain curve, via the 'blue shift' discussed in sec. 2.2.1. Resultingly, the main purpose of V_i is to shift the position of SOA gain curve in the wavelength domain.

Characterization of the Variable Couplers Implemented

Before implementing the couplers into the system, they are both characterized separately to confirm specification and document possible unexpected behavior. The detailed analysis is outside of the focus of this work. Here, a shortened repetition of the results is given [67].

Figure 3.6 shows the schematics of the setup used to characterize the variable couplers imple-

3.2. Introducing the Modifcations to Obtain a Functional ICAS Setup

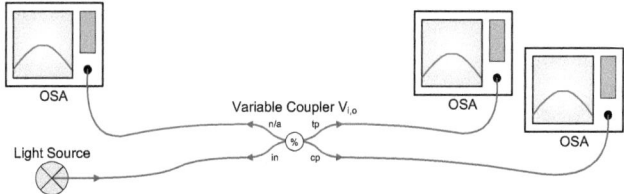

Figure 3.6.: Schematic of the setup used for the characterization of variable couplers $V_{i,o}$.

mented. It consists of a broad band light source[6], three OSA[7] and the variable couplers $V_{i,o}$ to be characterized. The couplers have four ports labeled in, tp, cp, and n/a. The in-port is to be connected to the light source, the tp- is the throughput port. cp denotes the coupled port where light is coupled out. The n/a port has no function if the light source is connected to the in-port. However, the variable couplers are symmetrical devices consisting of two intertwined fibers, one of them connecting in and tp, the other n/a and cp. Hence, all ports change their function, depending on where the light source is connected. In our case this is the in-port and for the opposite direction the tp-port. Their coupling ratio is adjusted be shifting the cores of the two fibers in respect to each other and thereby defining the amount of radiation to couple from one fiber to the other.

For both measurement directions light entering the coupler from the light source is divided to the three remaining ports, where the transmission spectra are recorded. The comparison of these spectra delivers information on coupling ratio for each wavelength in the entire wavelength range measured as well as the distribution of total optical Intensity $\mathscr{I}_{\text{int}}$, which is proportionate to

$$\mathscr{I}_{\text{int}} = \int_{1520}^{1570} \mathscr{I}_\lambda (\lambda) \, d\lambda, \tag{3.1}$$

where $\mathscr{I}_\lambda (\lambda)$ is Power Spectral Density.

Light in a laser resonator propagates in two directions. Hence, the variable couplers are characterized in two configurations:

- '→' with the configuration specified by the vendor, as shown in fig. 3.6.
- '←' with the light source connected to the tp-port.

---
[6] Agilent 83437A
[7] Agilent 86142B, Yokogawa AQ6370, Yokogawa AQ6375

3. Laser Setup Evolution

The results for a single pass measurements in comparison to vendor specification are shown in fig. 3.7 a and fig. 3.7 b.

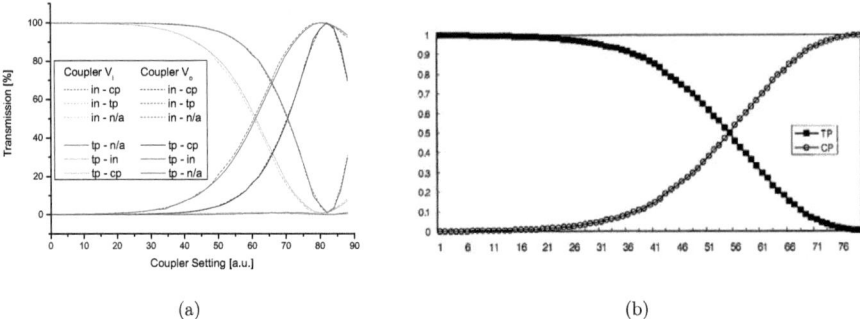

(a)   (b)

Figure 3.7.: (a) Measurements of coupling ratio of the variable couplers as a function of knob setting: the red curve denotes coupler V_i, blue curve coupler V_o.
(b) Specification provided by the vendor.

Increasing coupling ratio linearly according on the scale of a.u. initially results in a very small change in coupling ratio measured. A more pronounced change is observed from about 40 a.u. onwards. At about 60 a.u. coupling ratio approaches 50%. 100% coupling is reached at about 80 a.u..

In addition to vendor specification the measurement results presented in fig. 3.7 a include the unlabeled coupler port on the left, into which no light is coupled in an ideal case. Our measurements show that this does not apply to the couplers characterized. At 50% coupling ratio between the ports labeled tp and cp, the power measured at the port labeled n/a reaches a maximum of $(1.00 \pm 0.02)\%$. This behaviour is equally observed for both couplers in both configurations characterized. However, it can be treated as additional loss and has no negative effect on the functionality of the laser system. An important observation with these measurements is that both couplers do show different tuning response. The region of the scale, in which coupling response is strong differs between the two couplers. The same applies to the slope of the curves around 50% coupling ratio. This is very interesting especially when comparing the characteristics measured to vendor specification, which are equal for both couplers. As a consequence, the data measured is to be considered, when adjusting coupling ratio to a defined value, instead of relying on vendor specification alone.

Important characteristic values of the two variable couplers characterized are summed up in

3.2. Introducing the Modifcations to Obtain a Functional ICAS Setup

tab. 3.1. Detailed plots of all measurements can be found in [67].

Coupler V_i	'→'	'←'
Knob position @50% (a.u.)	(61 ± 1)	(61 ± 1)
max. loss @50% (%)	(1.0 ± 0.1)	(1.0 ± 0.1)
Coupler V_o	'→'	'←'
Knob position @50% (a.u.)	(71 ± 1)	(71 ± 1)
max. loss @50% (%)	(1.0 ± 0.1)	(1.0 ± 0.1)
Vendor Specification	'→'	'←'
Knob position @50% (a.u.)	55.5	55.5
max. loss @50% (%)	0	0

Table 3.1.: Characteristic values for the variable couplers $V_{i,o}$ characterized.

3.2.2. The Necessity for a Freely Propagating Beam

The previous work on the laser system by [48] focused mainly on its spectral specifications derived from the absorption characteristics of possible target substances for a sensor system basing on it. Oscillating wavelengths, the position of the gain curve and also the tuning mechanism have been specified and designed in accordance with a sensing application in mind. The final step, the implementation of a freely propagating beam, into which a sample can be inserted, is not yet implemented. The objective of this work is the proof of concept of the sensing principle described in patent DE 10 2004 037 519 B4 for which the implementation of a freely propagating section of the laser is an important step. This section deals with the collimator arrangement[8] implemented into the setup as the most convenient and promising approach for enhancing the setup by the possibility of introducing a sample. It is schematically shown in fig. 3.8.

Requirements to the Collimator Setup

If a sensor system is to reliably detect even smallest changes inside the resonator, the resonator itself including the collimator arrangement as part of it must be more stable than any change to be detected. Instability in the collimator arrangement implemented immediately transfer into a change of the optical properties and the resulting output of the system, which is evaluated as sensor response. Therefore, optical stability is the most important issue to be dealt with. Issues affecting this are mainly related to

---
[8]Thorlabs FFBM-S-1550-Y

3. Laser Setup Evolution

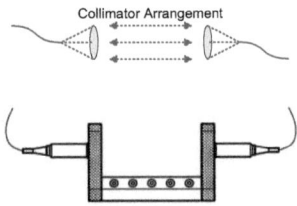

Figure 3.8.: Schematics of the collimator arrangement implemented.

- Changing temperature
- Mechanical stability
- Chemical stability
- Consideration of versatility

A change of temperature firstly means a change in the physical length as a result from thermal expansion and secondly a change of the refractive index in the collimated beam. Both affect the optical path length between the surfaces of the collimator lenses. This is delicate when regarding possible filter properties from interference effects of such a configuration.

Mechanical stability is critical for the same reason. Besides a change in path length, mechanical influences also affect the direction parallel to the optical axis. This means that also a misalignment by angular or lateral deflection of the collimated beam can reduce the intensity transmitted through such a collimator arrangement.

As the collimator arrangement is the component in the full laser setup, which is most prominently exposed to the substance to be analyzed, it has to be unaffected by the chemical properties of a sample. Potential target substances are solvents, aggressive acids or bases, which can potentially harm the optical components of the collimator arrangement.

Beside the stability issues mentioned above one important aspect is versatility towards sample structure and shape. Although a liquid sample is targeted at this stage, it is planned to exchange samples in future for solids or gases. For this reason the collimator arrangement must allow the introduction of any type of sample cell, or at least offer the possibility for enhancement in a way that different sample types can be inserted.

3.2. Introducing the Modifcations to Obtain a Functional ICAS Setup

Specification	FFBM-S-1550-Y
Wavelength	1550 ± 20 nm
Maximum Power	3 W
Insertion Loss	0.6 ± 0.3 nm
Return Loss	> 55 dB
Fiber	SMF-28e

Table 3.2.: Specification of the collimator arrangement characterized.

Collimator Arrangement Specification

The mechanical data of the collimation solution decided upon is shown in tab. 3.2. It consists of a solid base plate with ten holes of 3.3 mm in diameter, arranged on a 0.5" grid for mounting accessories. The actual collimation lenses[9] are held by two solid vertical pieces on the side. All parts consist of stainless steel, which has excellent material properties in respect to thermal expansion coefficient and chemical inertia. The collimator arrangement is mounted in thermal contact to the large steel plate of an optical table, which significantly increases thermal inertia and thereby reduces short term fluctuations in temperature.

For 1550 nm the beam diameter of the free space beam at the lens interface in the collimator arrangement is 0.38 mm with a divergence of 5.20 mrad. At these values the total insertion loss is specified with (0.6 ± 0.3) dB.

In order to confirm the compatibility with our application, the collimator arrangement is individually checked for compliance with specifications. The focus in this is laid on the independence from light propagation direction through the collimator arrangement, as its ports are surprisingly marked 'in' and 'out'. Measurements are obtained using the setup depicted in fig. 3.9 implementing the collimator arrangement between the light source[10] and OSA[11] in both directions possible.

The results of the measurements, which show the directional independence of insertion loss are summed up in tab. 3.3.

Due to difficulties with fiber splicing during the configuration of the setup, the relative errors obtained for insertion loss are quite large. Within the errors, however, no directional difference in the insertion loss of the collimator arrangement characterized is found.

---

[9] Thorlabs 352150-C Unmounted Geltech Aspheric Lens
[10] Agilent 83437A
[11] Agilent 86142B

3. Laser Setup Evolution

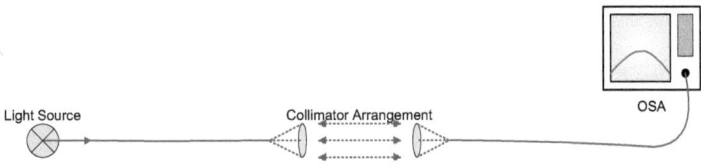

Figure 3.9.: Schematics of the setup used for the characterization of the collimator arrangement implemented.

	Insertion Loss
'IN' to 'OUT'	(0.4 ± 0.3) dB
'OUT' to 'IN'	(0.2 ± 0.2) dB
vendor specifications	(0.6 ± 0.3) dB

Table 3.3.: Optical properties of the collimator arrangement measured in two directions [67].

Collimator Arrangement Implementation

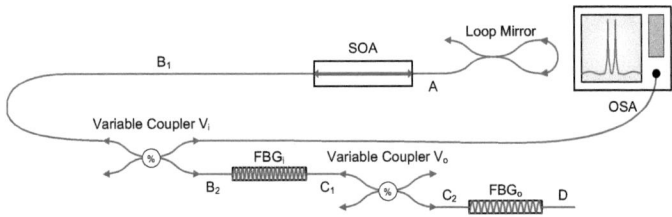

Figure 3.10.: Possible positions A-D for the implementation of the collimator arrangement.

Due to its open and flexible fiber based configuration, there are several different positions possible for the implementation of the collimator arrangement. In the following, the advantages and disadvantages of collimator arrangement placement in different positions are discussed. Figure 3.10 shows all possibilities for collimator arrangement placement in the laser system. For easy reference they are labeled from A, between loop mirror and SOA, to D, outside the cavity, behind both FBG.

- From a mode competition point of view positions C_{1+2} are very promising because absorption here only affects the mode belonging to M_o. M_i in C_i does not penetrate an

3.2. Introducing the Modifcations to Obtain a Functional ICAS Setup

absorber in the collimator arrangement and stays unaffected. As a result, it can be freely positioned very close to M_o regardless of the samples' absorption characteristics.

- Positions B_{1+2} and A are affected by both modes. Mode competition takes place, but the detection of absorbers requires absorption spectra that have significant changes in the tuning range of the FBG. From the mode competition point of view positions B_1 and B_2 mainly differ from a practical point of view, when thinking about modifying or further enhancing the setup, i.e. by exchanging the SOA, loop mirror or adding any parts to that end of the setup.

As already mentioned, positions C_{1+2} seem the most suitable positions for a collimator arrangement from the perspective of mode competition. Implementation here is discarded however, because at this point the adaptability of the setup to different samples is an essential criterion. Different samples require different sample chamber configurations, which also introduce different loss each. At present, the intensity tuning mechanism by reducing the reflectivity of FBG_o only allows equalizing mode intensity damping M_o relative to M_i. If sample chambers were to be introduced into the system that already suppress M_o below M_i, the intensity equilibrium, which is essential for the sensitivity of the sensor [52], can no longer be established by means of V_o.

As a result, position B_1 is chosen for the implementation of the collimator arrangement. In the final product, where the loss caused by the sample chamber is well known, position C_{1+2} will be reconsidered. The final setup used for the POC in chapter 5, including the collimator arrangement in position B_1, is shown in fig. 3.11.

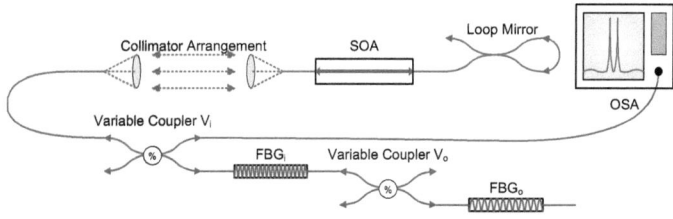

Figure 3.11.: Laser system including the collimator arrangement in position B_1.

61

3. Laser Setup Evolution

3.2.3. Thermal Instability in the FBG Feedback System

For the reliable detection of a target substance from defined absorption characteristics in its absorption spectra, quick, precise and stable addressing of these characteristics in the wavelength domain is inevitable. In conventional spectroscopes using a broad band light source this is mostly done by the filtering of broad band light by dispersive elements [25]. In the setup developed in this work, the narrow band emission is tuned to the wavelength desired by tunable FBG. FBG generally enable very stable reflection wavelength, as shown in [63]. In sec. 3.1.3 the FBG tuning mechanism is introduced. Here the focus lays on how output wavelength is further stabilized in comparison to the setup developed in [48].

Figure 3.12 shows the temporal development of FBG center wavelength after a change of FBG temperature. The measurement is performed by first setting the target temperature on the temperature controller[12], and then recording laser output in increments of 15 minutes after the moment when the temperature controller reaches its target temperature within tolerances of ±0.01°C. The plots obtained are generated by fitting Gaussian curves to the peaks in the spectra recorded and plotting their center wavelength versus time. The error bars are the standard deviation of center wavelength returned by the fit.

Five curves are shown: the black data points are FBG center wavelength λ with corresponding error. The dashed black line is a second order exponential fit, eq. 3.2, applied to this data, referred to as λ_{fit}. The change $\Delta\lambda_{\text{fit}} = \lambda_{\text{fit}}(n+1) - \lambda_{\text{fit}}(n)$ of fitted wavelength between the recent and previous reading is shown as a solid blue curve. It can be interpreted as the drift of FBG center wavelength from thermal system inertia. The dashed blue line shows the mean of the absolute deviation $\delta\lambda$ of data from the fit. It can be regarded as the magnitude of wavelength fluctuations from environmental influence. The red curve is the sum of $\delta\lambda$ and $\Delta\lambda_{\text{fit}}$ and indicates the total fluctuation of FBG center wavelength.

$$\lambda = y_0 + f_2 \cdot \exp^{-t/f_3} + f_4 \cdot \exp^{-t/f_5}, \qquad (3.2)$$

where t and λ are independent and dependent variable time and wavelength, respectively and y_0 and $f_1 - f_5$ are free fitting parameters.

The measurement data show that initially the thermal drift of the system is still strong. After about 300 min however, FBG center wavelength is stable within tolerances of ±0.3 pm.

Two time regimes of wavelength stability are identified from optical output characteristics:

---
[12]Eurotherm 3216

3.2. Introducing the Modifcations to Obtain a Functional ICAS Setup

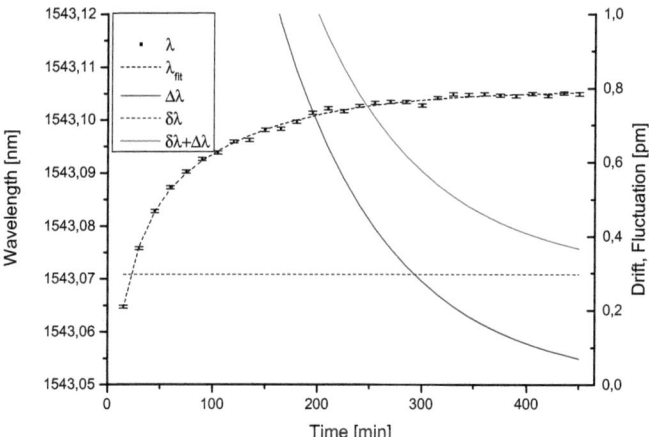

Figure 3.12.: Stabilization behavior of the FBG feedback system in the initial laser system by [48]. Final stability is ultimately limited by thermal fluctuation of the environment.

3. Laser Setup Evolution

1. $\Delta\lambda_{\text{fit}} > \delta\lambda$: the thermal inertia of the system dominates wavelength stability

2. $\Delta\lambda_{\text{fit}} < \delta\lambda$: environmental influences dominate wavelength stability

In the regime where the thermal inertia of the system dominates wavelength stability mainly the responsivity of the PID[13] temperature controller and the heating power of the heaters in relation to the thermal inertia of the aluminum block play a role. PID parameters and temperature controller responsivity are optimized by software, which is implemented into the temperature controllers.

The final temperature stability in fig. 3.12 is mainly dependent on environmental influence on the FBG. Assuming an ideal environment, the system eventually reaches a perfectly stable thermodynamic equilibrium. In this case no thermal fluctuations are observed and FBG center wavelength remains stable. Thermal fluctuations in the environment or convection however mean distortions from this ideal case, resulting in small permanent fluctuation in FBG center wavelength. FBG center wavelength remains stable. Thermal fluctuations in the environment or convection, however, mean distortions from this ideal case, resulting in small permanent fluctuation in FBG center wavelength.

The point in time, when $\Delta\lambda_{\text{fit}} = \delta\lambda$ is defined as the stabilization time $t_{\text{stab.}}$ of the FBG system, after which no significant further stabilization is expected as environmental influences become the main stability defining factor.

Possibilities of Optimization

There are two ways in which the final stability of the FBG system can be improved: minimizing the thermal inertia of the system to achieve shorter settling times and reducing environmental influence to achieve better final stability.

Minimizing the thermal inertia of the system involves a redesign of the tuning mechanism in terms of power, material and layout as system drift mainly depends on these parameters. It results in a more rapid approach of thermal equilibrium and thereby the final value of λ. As a result the blue curve in fig. 3.12 is expected to change to a more square shape, which leads to a reduction of total fluctuation of the system for $t < t_{\text{stab.}}$. In the time regime of $t \gg t_{\text{stab.}}$ no change in total fluctuation of the system is expected as in this time regime $\Delta\lambda$ is insignificant anyhow.

---

[13]Proportional Integral Differential

3.2. Introducing the Modifcations to Obtain a Functional ICAS Setup

A reduction of the environmental influence on the FBG system targets the final stability of the system in the time regime $t \gg t_\text{stab.}$. It can be achieved by thermally uncoupling the FBG system from the environment. In fig. 3.12 this uncoupling results in a shift of the dashed blue line to smaller values. Although this means shifting $t_\text{stab.}$ to larger values, the final value of the sum of $\Delta\lambda$ and $\delta\lambda$ in the regime of $t \gg t_\text{stab.}$ is smaller.

In this work the latter approach is pursued, as stabilization time is not critical at the present stage. The final stability however is of great importance, as it influences all calculations based on FBG wavelength. Thermal uncoupling is implemented by Styrofoam enclosures for the FBG system. Their impact on wavelength stability is investigated in detail in [67]. The results relevant for the POC are repeated in sec. 4.1.6.

3.2.4. The Elimination of Parasitic Cavities

A laser cavity is an arrangement of optical components that confines EM-radiation of specific wavelength. For the simple FP laser introduced in sec. 2.2.1 this is achieved by reflection of radiation at the resonator mirror surface, whose distance defines the modes supported. In such an ideal FP laser these mirrors are the only reflecting surfaces and hence its output spectra perfectly reflect the FP condition given in eq. 2.7.

In a real laser system, however, this is not the case: imperfections in the laser setup like imperfect AR coatings on resonator components, slight variations in refractive index along the cavity or bad interconnections between components cause partial reflection of the radiation confined in the resonator. These reflections also form weak cavities, which are noticeable as additional features in the output spectra of a laser system. Their effect is mostly undesired and contributes to the noise of the system. Such undesired cavities are called parasitic cavities. In the laser system developed in this work many sources for parasitic cavities are present.

A good example of one is shown in fig. 3.13, the output spectra of the ASE from the SOA implemented. For this measurement the SOA is isolated and connected directly to an OSA[14] under forward bias. Ideally a smooth bell shaped curve is expected, directly resembling SOA gain. Figure 3.13, however, inhibits a superstructure on top of the bell shaped curve, which results from a parasitic cavity effect. At closer look it results from two different cavities [48]. One consists of the two imperfectly AR coated facettes of the semiconductor crystal, and another one is formed between semiconductor crystal and the ends of the fibers attached.

---
[14] Agilent 86142B

3. Laser Setup Evolution

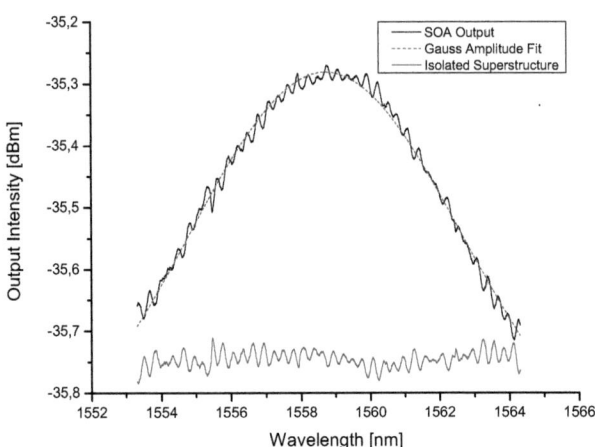

Figure 3.13.: Measured spectral output of the SOA. A superstructure is observed on the broad Gaussian gain curve.

The complete laser system described in [48] consisting of SOA, FBG, coupler and loop mirror contains many more sources for parasitic cavities than the SOA alone regarded in fig. 3.13. It is far too complex for a detailed analytical investigation of parasitic cavities and their effect on output spectrum due to the large number of components implemented. The general approach of removing all sources for unintended reflection is pursued, which is expected to reduce the formation of their associated parasitic cavities. As all commercially purchased parts are already AR coated, housed and can hence not be further optimized, the efforts described in this work focus on reducing the reflectivity from Fresnel reflexes (eq. 2.15) of the fiber butts in the laser system by implementing optical isolators[15]. In the following, their working principle is explained and the characterization of four isolators implemented into the laser system presented.

Isolators

What a diode does with electric current, isolators do for light. Optical isolators transmit light unhindered in forward direction, but not in reverse, implemented by exploiting the polarization properties of light. In its most simple form an isolator consists of the three components shown

---
[15]Thorlabs M-II-1-15-H-L-E-1

3.2. Introducing the Modifcations to Obtain a Functional ICAS Setup

in fig. 3.14 [55]:

1. A polarizer
2. A Faraday rotator, rotating the plane of polarization by an angle of 45°
3. A polarizer oriented 45° in respect to the first

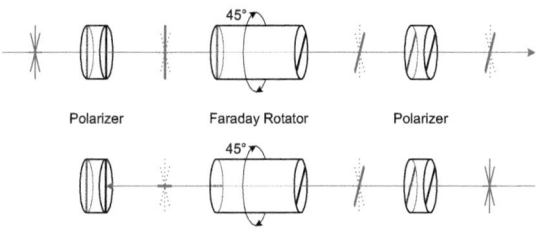

Figure 3.14.: Schematic illustration of the working principle of a polarization dependent optical isolator.

In forward direction the incident light is firstly polarized linearly by the polarizer in fig. 3.14. A Faraday rotator then rotates the plane of polarization by 45° in respect to the first polarizer so it can pass the second polarizer unattenuated.

In reverse direction, the sequence is vice versa: the incident light is firstly polarized by the polarizer on the right. As the Faraday effect is not reversible, the Faraday rotator rotates the plane of polarization by 45° in the same direction as in the case of forward direction. Now it is polarized by an angle of 90° in respect to the polarizer on the left, which it hence cannot pass.

This simple schematic only works for linearly polarized light. A similar concept for unpolarized light, which bases on splitting incident light by polarization plane, is described in [81]. The isolators implemented in this work are specified for unpolarized incident light.

Validation of Specifications

Before implementation into the laser system, all isolators purchased are tested for validity of their specification, shown in tab. 3.4. Special point of interest is isolation characteristics at $\lambda = (1543 \pm 2)$ nm, as the laser system developed in this work is designed for that wavelength range.

3. Laser Setup Evolution

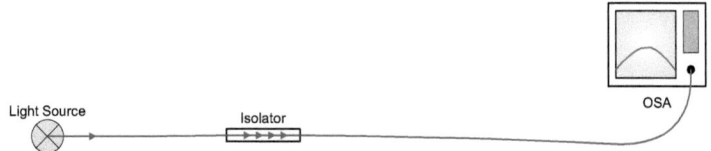

Figure 3.15.: Schematics of the setup used for the characterization of optical isolators.

For verification of specification the isolators are connected in forward and reverse direction, respectively, between a broad band light source[16] and an OSA[17], as shown in fig. 3.15. Beforehand the spectrum of the broad band source is recorded as a reference.

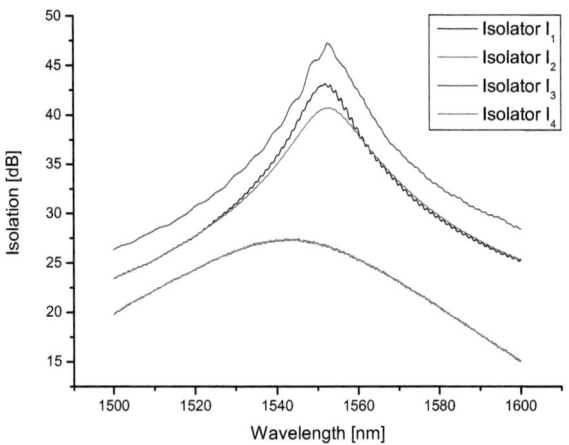

Figure 3.16.: Measured spectral throughput of the isolators measured in reverse direction.

In forward direction no wavelength dependent behavior is observed. Insertion loss L_{fw} is white and in fair compliance with vendor's specifications. The spectra obtained in reverse direction are shown in fig. 3.16. Isolation Ξ_{max} is highly wavelength dependent in the range measured and exhibits a maximum around 1550 nm for isolators I_{1-3}. Ξ is >40 dB for 1550 nm for isolators I_{1-3}, and still > 35 dB at 1542 nm, the wavelength interesting for the laser system

---

[16]Agilent 83437A
[17]Agilent 86142B

3.2. Introducing the Modifcations to Obtain a Functional ICAS Setup

	Isolator I_1	Isolator I_2	Isolator I_3	Isolator I_4
spec. L_{fw}/dB	0.29	0.36	0.58	0.58
meas. L_{fw}/dB	(0.74 ± 0.01)	(0.32 ± 0.01)	(1.68 ± 0.02)	(0.33 ± 0.01)
spec. Ξ_{max}/dB	43.13	40.71	47.25	27.51
$\Xi_{1542\,nm}$/dB	(37.1 ± 0.3)	(36.1 ± 0.2)	(39.7 ± 0.2)	(27.3 ± 0.1)
λ @Ξ_{max}/nm	(1551.8 ± 0.1)	(1552.7 ± 0.1)	(1552.6 ± 0.1)	(1542.3 ± 0.1)

Table 3.4.: Specification and measurements of the isolators $_{1-4}$ tested.

developed in this work. Isolator I_4 seems to have some kind of defect. Its isolation maximum is shifted and maximum isolation is significantly lower than for isolators I_{1-3}. Consequently, it is implemented into the laser system in the least critical position with its limited capabilities in mind. Although isolators I_{1-3} inhibit better characteristics than Isolator I_4 they are still far away from an ideal device with completely wavelength independent isolation and no insertion loss. This, however, is not a problem for the intended application. The exact parameters for all four isolators tested, obtained from Lorentzian approximation of the spectra measured, are summed up in tab. 3.4.

Implementation into Laser System

The setup including the isolators, which are tested in this section is shown in fig. 3.17. The isolators are labeled in the same manner as in tab. 3.4. Isolator I_4, which inhibits rather bad properties in testing, is placed at the end of the fiber system, the farthest away from the SOA. The impact of the parasitic cavity effect caused by this fiber butt is quite small due to its position behind both variable couplers. Implementing isolator I_4 here despite its deficiency can be argued, as it leaves the well performing isolators for the fiber butts with stronger influence on the laser system. The other fiber butts are equipped with isolators I_{1-3}.

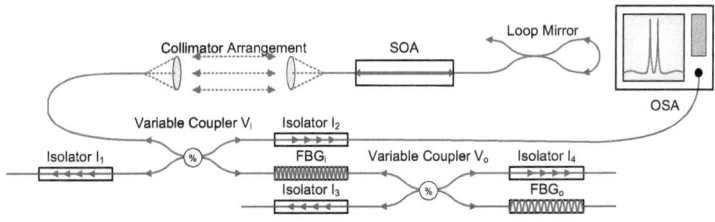

Figure 3.17.: Laser system including isolators, numbered according to tab. 3.4.

3. Laser Setup Evolution

3.2.5. Reducing Linewidth

For high wavelength resolution of a sensor system based on the two mode laser system dealt with in this work, the line width of the individual modes is a critical parameter. As prerequisite to take the total intensity of one of the modes as a measure for the absorption of a substance, its $FWHM_{mode}$ has to be smaller than the bandwidth of the absorption line to be measured. This condition ultimately limits the resolution of such a sensor system by design. Consequently, small mode line width is very desirable for high resolution.

In its initial state, the system's $FWHM_{mode}$ is $0.07 - 0.08$ nm, which is already sufficiently small to detect the absorption characteristics of liquids.

However, a precise quantification of details, such as the exact slope or curvature on the side of broad absorption peaks gains in accuracy with decreasing mode line width. As one of several improvements of the system in this work, bad splicings are identified as a line width broadening factor. The following section deals with their replacement.

Splicing Loss

In determining the attenuation caused by imperfect splices in the initial laser system the setup depicted in fig. 3.18 is used. It consists of a broad band light source[18] and an OSA[19] for spectra recording. Between a fiber is implemented, which is repeatedly cut and re-spliced. Finally the attenuation per splice is calculated from a number of 15 splices in total. Measurements performed in [67] return an insertion loss L_{ins} of (0.3 ± 0.2) dB for an average single splice.

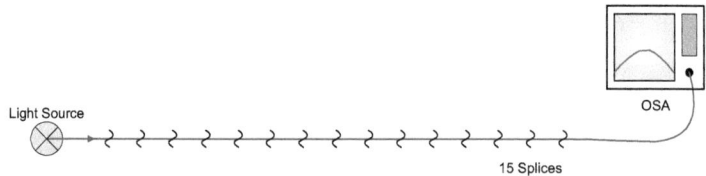

Figure 3.18.: Schematics of the setup used for the measurement of splicing loss.

This is a very high value compared to the 0.01 dB mentioned as typical attenuation for fiber to fiber splices in literature [60]. Resultingly, the laser system is disassembled and rebuilt, using

---
[18]Agilent 83437A
[19]Agilent 86142B

3.2. Introducing the Modifcations to Obtain a Functional ICAS Setup

rented splicing equipment[20]. During this process the attenuation on each splice is measured in-situ and documented. Figure 3.19 shows the final version of the laser system with all splices marked. Table 3.5 lists the attenuation measured for each individual splice.

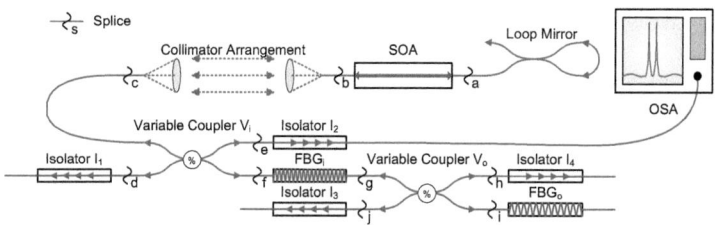

Figure 3.19.: Laser system including all refined splices a-j.

Position	L_{ins}/dB
a	0.01
b	0.01
c	> 0.01
d	0.03
e	0.01
f	0.03
g	0.01
h	0.01
i	0.02
j	0.01

Table 3.5.: Measured optical loss of the individual splices a-j.

The reduction of splicing loss is impressive. Where a typical splice causes an attenuation of about 0.3 dB in the initial laser system from [48], the typical loss introduced by a splice made with state of the art splicing equipment is only about 0.01 dB. Considering the total number of 11 splices in the laser system this improvement of splicing loss results in a reduction of total system loss by > 3 dB, also reducing the number of sources for parasitic cavities significantly.

---

[20]Fiberoptic-Solution OptiSplice III LID

3. Laser Setup Evolution

3.3. Liquid Cell Implementation

The POC, which is the main objective of this work, focuses on Propofol, a narcotic. This work focuses on detecting it in liquid phase, as the liquid provides pronounced characteristic absorption around 1550 nm. Details about the substance as such are presented in sec. 4.2.1.

This section deals with the requirements that liquid Propofol defines for the construction of a sample cell in which it can be introduced into the laser system. Firstly, the optical prerequisites for a sample cell are discussed and the cells decided for presented. Subsequently, the design of the sample holder constructed for the POC measurements is shown and explained.

3.3.1. Optical Prerequisites of the Cells Used

As the laser sensor system investigated in this work bases on mode competition between M_i and M_o defined by the reflectivity of FBG, it is important that sample cells do not influence the resonator's optical characteristics in favor of either one. In simple terms this means that the transmission of a cell implemented must not be wavelength selective in the wavelength range of the laser system. The same is also valid for A, R, S, as shown in eq. 3.3.

The transmission T of normalized white incident radiation through a piece of glass can be expressed as

$$T = 1 - A - R - S, \qquad (3.3)$$

where A is absorption, R represents reflection and S scattering. R is a parameter that involves material properties. A and S are additionally related to manufacturing quality.

- Absorption in the glass mainly results from contamination with absorptive compounds, such as hydrogen or hydrogen compounds. OH-compounds in particular are the most prominent example of an optically problematic contamination, especially in telecommunication fibers, where their reduction has been subject to intensive research [92].

- Scattering mostly happens at impurities, such as enclosures of air bubbles or particles.

- Surface reflectivity is given by the Fresnel eq. 2.15 and depends solely on refractive index contrast between the glass and its surrounding material.

3.3. Liquid Cell Implementation

All factors mentioned above have to be considered in the selection of material, geometry and implementation of sample cells in order to obtain flat transmission characteristics.

Apart from material properties an additional effect, which taken into account when considering the optical properties of a sample cell is possible interference, like the parasitic cavities mentioned in sec. 3.2.4.

3.3.2. Cell Material and Design

The commercially available glass with low absorption that is commonly used for tasks in optical spectroscopy from Ultra Violet (UV) up to NIR, which is chosen is fused silica glass[21], which is highly transparent and very stable under exposure to laser radiation. A typical transmission spectrum of this material obtained from [37] is shown in fig. 3.20 a. The plot shows reduced transmission around 1400 nm. As shown in fig. 3.20 b, it is very flat in the wavelength range from 1540 to 1550 nm and well suited for application in the laser sensor system designed in this work.

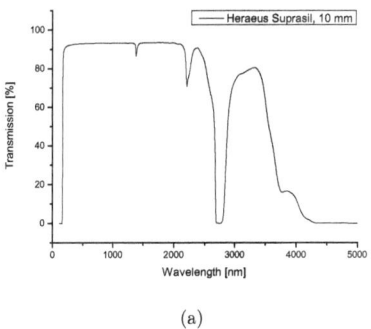
(a)

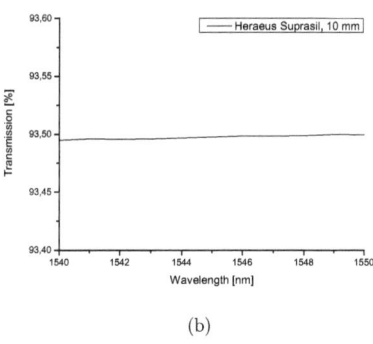
(b)

Figure 3.20.: (a) Broad band transmission spectrum and
(b) Narrow band transmission spectrum of 10 mm Suprasil provided by the vendor [37].

Plane parallel plates and commercial cells for absorption measurements made from fused silica are commercially available in different sizes and geometries[22], [23]. The four cells purchased are

---
[21]Heraeus Suprasil
[22]LINOS Photonics GmbH & Co. KG
[23]Hellma GmbH & Co. KG

3. Laser Setup Evolution

Hellma cells 110-QS	1 mm	2 mm	5 mm	10 mm
Volume	350 µl	700 µl	1750 µl	3500 µl
Outer Dimensions				
Height	52 mm			
Width	12.5 mm			
Depth	3.5 mm	4.5 mm	7.5 mm	12.5 mm
Inner Dimensions				
Width	9.5 mm			
Base Thickness	1.5 mm			
Number of windows	colspan 2			
Optical Properties				
n_{eff} @1550 nm	1.44			
OH-Concentration	< 1000 ppm			

Table 3.6.: Vendor specification of the sample cells purchased.

shown in fig. 3.21 a, their specifications in tab. 3.6.

All cells in tab. 3.6 are simulated for filter properties by an open source software [53], which bases on the transfer matrix method described in [73]. The simulation results of different configurations of the 10 mm cell are shown in fig. 3.21 b. The cells of thickness 1, 2 and 5 mm behave similarly. The only difference between the various cell thicknesses is the spacing of the parasitic cavity modes, which behaves reciprocal to optical cell length and is hence larger for smaller cells.

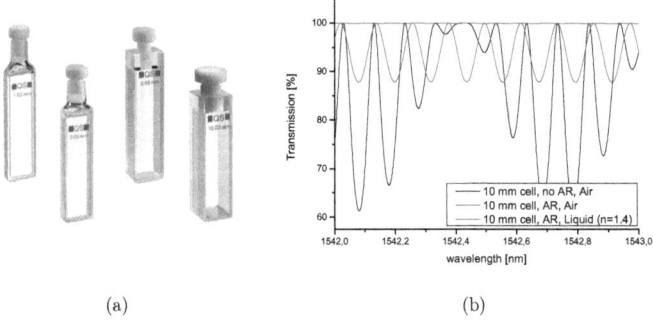

(a)   (b)

Figure 3.21.: (a) Sample cells purchased.
(b) Simulation of the 10 mm sample cell's spectral transmission.

The curves in fig. 3.21 b show the parasitic cavity mode structure resulting from the 10 mm cell for laser radiation of 1550 nm at normal incidence. Parasitic cavity modes are most pronounced for the empty cell without AR coating. The application of AR coating to the outer interfaces eliminates the mode structure resulting from these interfaces, leaving only the characteristics of the inside interfaces behind. These, however undesirable they may be, cannot be further reduced by the straight forward approach of applying AR coating for technological reasons:

- On the inside of a pre-assembled cell, they are unreachable by the plasma used for layer deposition.
- Coating before cell assembly is not possible because any coating is destroyed in the assembly process.
- The most important reason, however, is the immediate contact an AR coating on the inside of such a sample cell has to the substance to be analyzed. As the manufacturers do not disclose any information on the exact composition of their AR coatings, it cannot be risked exposing the samples to them, as chemical reactions, which might significantly change a sample's properties, cannot be excluded and avoided.

It also has to be noted that these simulations account for very strictly monochromatic light, with an ideal FWHM of zero. In that case the very clear mode structure seen in fig. 3.21 b is expected, which is a major difference to the real laser system. However, the finite mode FWHM of the actual laser modes M_i and M_o results in much less pronounced interference structure in reality. An additional reduction of parasitic cavity impact is achieved by mounting the sample cell at an angle of $1 - 2°$ in respect to normal incidence. This way reflections are directed out of the resonator and cannot be coupled back into it, where they are amplified by the SOA.

As a general solution to the problem of interference, a more refined measurement technique is chosen, which bases on measuring the absorption of Propofol in a solution of solvents of similar refractive index to that of fused silica; for details see sec. 4.2. This way, the refractive index contrast between fused silica and the inside of the sample chamber is reduced significantly, decreasing the impact of corresponding parasitic cavity modes. The simulation for the 10 mm cell filled with the index matching liquid is also included in fig. 3.21 b.

3.3.3. The Design of a Sample Cell Holder

To introduce the sample cell into the laser system in a reproducible and mechanically stable way, a holder is designed that fits into the collimator arrangement described in sec. 3.2.2. In the

3. Laser Setup Evolution

design of the holder special attention is paid for reproducible mounting, mechanical stability and adjustability for the different cell dimensions. It allows to adjust the angle of incidence for the sample cells to not re-couple reflections back into the laser system and compensates for any beam offset caused by this angled mounting. This section presents the design and features of the sample holder in detail.

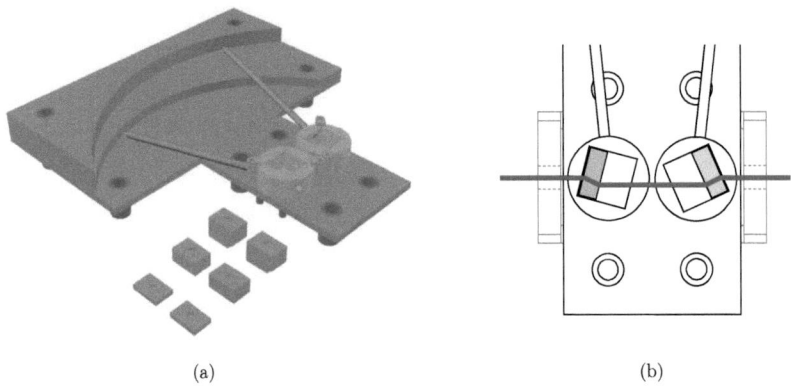

Figure 3.22.: (a) Complete sample cell holder.
(b) Schematics of the optical path through the holder including sample cell and beam offset compensation.

Figure 3.22 a shows the holder with all its components designed to fit into the collimator arrangement described in sec. 3.2.2. The individual parts in the schematic are color coded by functionality.

Blue specifies the parts that belong to the Aluminum mounting plate. It serves a a base for all other parts contained in the design and is firmly attached to the underlying optical table (not shown) by eight 6 mm screws. The holes hold cylindrical brass rods, colored in a darker blue, aligned on a 25 mm grid equal to that on the optical table and define the mounting plate's height above it. On its bottom, in the area inside the collimator arrangement, the mounting plate has 6 more smaller cylinders that fit snugly into mounting holes in the collimator arrangement, holding that in a defined position as well. Additionally to serving as a base, its top region also serves as a scale for adjustment of the angle between the optical interfaces of the cell and the optical axis of the laser system.

The parts holding and defining the position of the actual sample cells are colored in **gray**. The light gray colored cylinders, milled from brass and eloxated in black, hold the optics. They rest

3.3. Liquid Cell Implementation

in round holes in the mounting plate, in which they rotate freely. The indicators of a length of 80 mm and 100 mm extending from the left and right cylinder, respectively, are used to show the angle between the optical interfaces of the components held by the holder and the optical axis of the collimator arrangement.

Red identifies brass clamps, used to hold the rotatable gray sample holders in place. They are attached to the mounting plate by M2 screws.

The **green** blocks at the bottom left in fig. 3.22 a show spacers to be inserted into the square holes in the gray cylindrical sample holders, which compensate for the different thicknesses of the components to be mounted. They are held in place by small screws. These parts are made of Teflon, which is soft enough to not damage the fused silica cells, and also does not react with Propofol or solvents.

Functionality explained

The optical path through the collimator arrangement including the sample holder with sample cells is schematically shown in fig. 3.22 b. Here radiation is considered entering the first collimator lens on the left and propagating to the right.

Laser radiation entering from the fiber system is expanded by the collimation lens and propagates freely along the optical axis of the collimator arrangement. The left cylindrical holder retains the sample cell, where the laser radiation is subject to wavelength selective absorption. The cell is mounted at an angle of $1-2°$ to avoid re-coupling reflected radiation back into the resonator. Additionally, due to the angled mounting, the collimated laser experiences a parallel offset from the optical axis of the collimator arrangement, dependent on sample cell thickness and mounting angle. The second cylindrical holder in the collimation system holds a plate of fused silica, also mounted at a defined angle in respect to the optical axis of the system. It compensates the laser beam offset from the optical axis of the system obtained by the angled optical interfaces of the sample cell.

Using the device

Introducing cells or exchanging cells against others of different thicknesses requires a few steps to be carried out in correct order to ensure optimum operating conditions of the laser system:

1. The laser system is switched off.

3. Laser Setup Evolution

2. The screws in the cylindrical holders marked in gray in fig. 3.22 a are loosened and a sample cell inserted into one of the holders along with corresponding spacers.
3. The other holder is equipped with an AR coated plate of fused silica and its corresponding spacer.
4. The Teflon screws are tightened.
5. The desired angle between sample cell and optical axis is set by the indicator attached to the cylinder holding the cell.
6. To realign the laser beam with the optical axis of the collimator arrangement, the laser system is switched on and the total power observed on an measurement device attached.
7. During laser operation at constant parameters, the indicator of the cylindrical holder containing the glass plate is moved. This movement can be observed as a change in the resulting power on the power meter or OSA. The correct position of the indicator for ideal beam offset compensation is found when the power on the measurement device is at maximum.

The collimator arrangement is now ready for absorption measurements.

3.4. Final Setup for the Proof of Concept

Figure 3.23 shows the laser system setup after the enhancements and optimizations presented in sec. 3.2 (a) in comparison to the initial laser system from [48] (b). The most obvious difference is that the enhanced setup is a lot more complex than the initial one. The single fixed ratio coupler is replaced by a variable one and an additional variable coupler is introduced between the two FBG. The first one allows varying broad band loss in the system and thereby manipulating the threshold conditions; the latter enables equalization of mode intensity for almost any position of M_i and M_o.

Where the laser radiation is guided only in fibers in the initial setup, the enhanced setup features a collimator arrangement with free beam propagation into which samples of various forms and compositions can be inserted. For POC measurements on liquids the enhanced setup also includes a cell holder that can hold commercial cells of thicknesses up to 10 mm. Besides the implementation of new functionality also the overall performance of the laser system has been improved in respect to thermal, optical and mechanical stability as well as line width and parasitic cavity effects.

3.4. Final Setup for the Proof of Concept

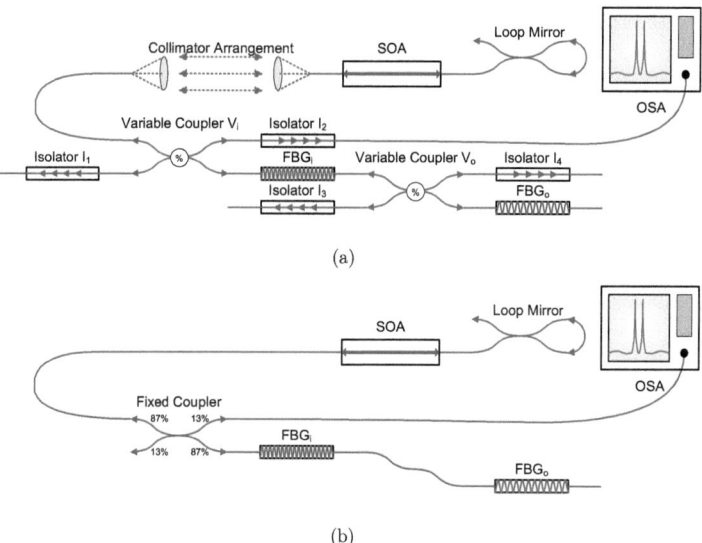

Figure 3.23.: (a) Enhanced laser system developed in this work in comparison to (b) the laser system from [48].

4. Experimental Work

In order to conduct reliable measurements on target substances, the setup depicted in fig. 4.1 developed in this work is characterized in respect to its optical properties and their dependence on the various input variables such as loss introduced by the variable couplers, gain blue shift from change in the injection current applied to the SOA and gain redshift from temperature of the SOA. Also the tuning response of the oscillating modes is characterized, so that defined wavelengths can be addressed and maintained during the measurement of the absorption of liquid samples. This chapter is divided into two parts:

- The characterization of the enhanced laser system
- The characterization of Propofol and the solvents used

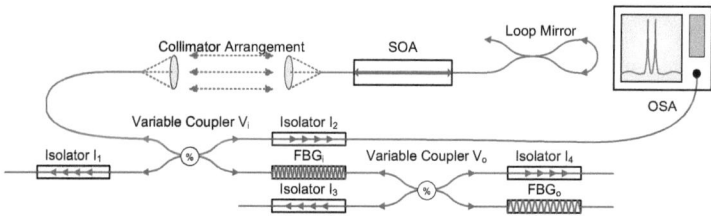

Figure 4.1.: The laser system developed in this work.

4.1. Characterization of the Enhanced Laser System

4.1.1. Analyzing Laser Output

Figure 4.2 a and fig. 4.2 b show exemplary laser system output. In fig. 4.2 a a broad band view on the full output spectrum is depicted. Its main features are the broad band ASE, which

4. Experimental Work

is equivalent to the gain curve of the SOA with its maximum at 1578.6 nm and a FWHM of 40.7 nm. Both, the position of the maximum and also the FWHM of the gain curve, are dependent on injection current and operating temperature of the SOA. These dependencies are characterized in detail in sec. 4.1.2 and sec. 4.1.2, respectively.

Additionally to the broad gain curve, the two laser lines corresponding to the wavelength positions of the FBG are visible in fig. 4.2 a. As the wavelength range chosen in this plot is too large to resolve the individual modes, a blow up of this area is shown in fig. 4.2 a.

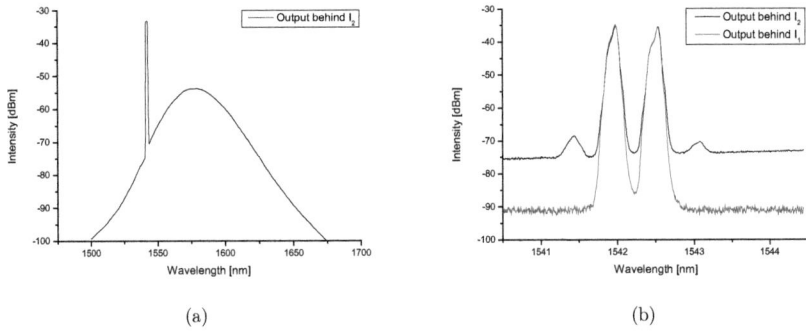

Figure 4.2.: (a) Measurement of the broad band output of the enhanced laser system at $I_{inj} = 113.6$ mA and $\Theta_{SOA} = 25°C$.
(b) Detail of the two individual laser modes at $\lambda_i = 1542$ nm and $\lambda_o = 1542.5$ nm.

In fig. 4.2 b two curves are shown: the black one shows laser output taken behind Isolator I_2 in fig. 4.1. The red curve is taken behind Isolator I_1. Hence, the black curve contains ASE from the SOA, which is responsible for the high noise level of this curve. The red curve only contains the laser lines returned from the FBG; noise level is significantly lower here. The back reflection of any ASE at any of the setup's facets is suppressed by the isolators in the setup and therefore only visible in one propagation direction.

The laser lines in fig. 4.2 b are located at $\lambda_i = 1542$ nm and $\lambda_o = 1542.5$ nm and have a FWHM of typically <0.06 nm, which varies with injection current and SOA temperature. The Q-factor calculated from the spectrum shown in fig. 4.2 b by eq. 2.11 equals 25700. SNR is greater than 50 dB. This value is on the level of state of the art telecommunication systems [33].

The exact dependence of FWHM and SNR on injection current and SOA temperature are discussed in sec. 4.1.4.

4.1.2. The Effect of Temperature and Injection Current on the Gain Profile

Blue Shift from Current

To characterize the behavior of the gain profile of the laser system under variation of injection current the OSA is connected to the setup in fig. 4.1 behind Isolator I_2. The two lasing modes are not important in this measurement, so they are prevented from reaching threshold by coupling all light out of the system at the variable coupler V_i. This fully suppresses feedback from the FBG behind it, keeping the laser system unaffected by lasing on the FBG wavelengths. Readings are taken by setting the SOA to a defined temperature and waiting for stabilization. Then injection current is increased in increments of 5 mA, and recording the optical output averaged over 10 readings on the OSA.

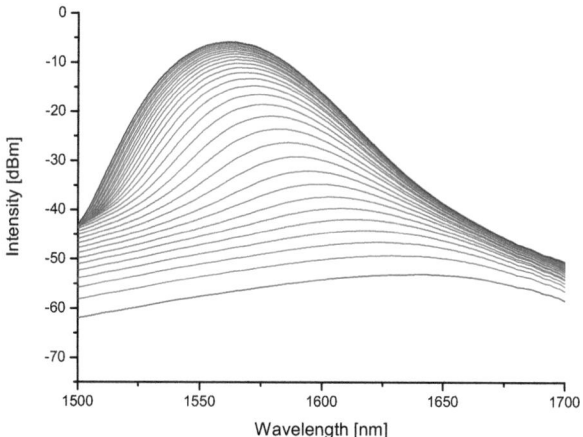

Figure 4.3.: Measurements on SOA gain curve at $\Theta_{\text{SOA}} = 15°C$ for injection currents from $I_{\text{inj}} = 5 - 150$ mA.

Figure 4.3 shows the output of the system for injection currents of $5 - 150$ mA in increments of 5 mA for $\Theta_{\text{SOA}} = 15°C$. The bottom curve represents the output at 5 mA, the top curve shows output at 150 mA; the curves in between correspond to injection currents of $10 - 145$ mA, respectively. The data plotted for $\Theta_{\text{SOA}} = 15°C$ has also been recorded for $\Theta_{\text{SOA}} = 20°C$, 25°C

4. Experimental Work

and 30°C but is not shown here for simplicity.

Two major tendencies are observed with increasing current, which are expected from theory:

1. The shift of the gain curve's maximum towards shorter wavelengths.
2. The reduction of gain FWHM from the increased amplification that affects the wavelengths closest to the maximum most.

In order to quantify these trends a more detailed analysis of the data is required. First, all curves are approximated by a Gaussian Amplitude fit, eq. 4.1

$$y = y_0 + f_1 \cdot e^{-(x-x_c)^2/2f_2^2}, \qquad (4.1)$$

where x and y are the independent and dependent variable, respectively and y_0, x_c and f_1, f_2 free fitting parameters. From the fit functions obtained the center wavelength λ_c and FWHM is calculated.

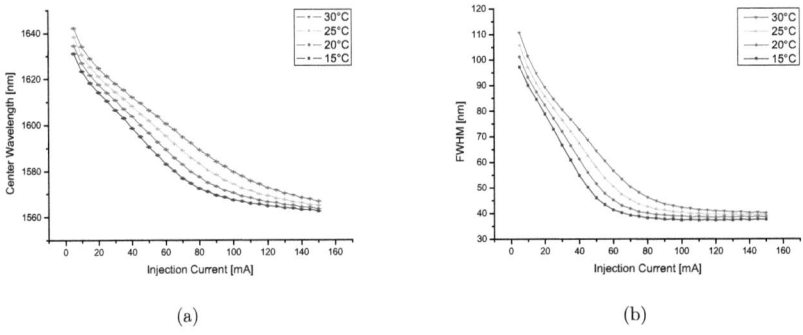

(a) \hspace{4cm} (b)

Figure 4.4.: (a) Measurements on the center wavelength of the gain curve of the SOA at $\Theta_{\text{SOA}} = $ 15, 20, 25 and 30°C for injection currents from $I_{\text{inj}} = 5 - 150$ mA.
(b) Measurements on the FWHM of the gain curve of the SOA at $\Theta_{\text{SOA}} = $ 15, 20, 25 and 30°C for injection currents from $I_{\text{inj}} = 5 - 150$ mA.

Figure 4.4a shows the center wavelength of the gain curve for $\Theta_{\text{SOA}} = $ 15, 20, 25 and 30°C and currents from $5 - 150$ mA. For all four temperatures measured, the maximum wavelength is varied by about 65 nm via injection current in the current range between 5 and 150 mA. The response is not strictly linear; however, as a rule of thumb a tuning response of 0.5 nm/mA can be presumed.

4.1. Characterization of the Enhanced Laser System

A plot of the behavior of the FWHM of the gain curve is presented in fig. 4.4 b. The data also originates from the Gaussian amplitude function fitted by eq. 4.1. The FWHM also decreases with increasing injection current. At about 80 mA it stabilizes at a value of 40 − 45 nm depending on temperature.

Gain Red Shift from Temperature

The characterization of the temperature redshift of the gain curve is performed analogous to the characterization of the injection current dependent blue shift described in sec. 4.1.2. The measurement setup and position of taking readings are the same. The spectra are obtained by setting injection current to defined values of 80, 100, 120 and 140 mA and increasing SOA temperature from 5 to 30°C in increments of 1°C for each injection current setting. For each combination of injection current and SOA temperature the output spectra of the system are recorded, averaged over 10 readings.

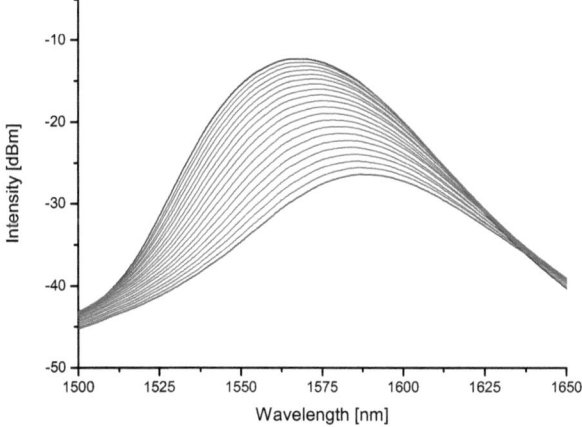

Figure 4.5.: Measurements on the SOA gain curve at $I_{\text{inj}} = 80$ mA for SOA temperatures from $\Theta_{\text{SOA}} = 10 - 30°C$.

Analog to above, fig. 4.5 shows the output of the system for SOA temperatures of $\Theta_{\text{SOA}} = 10 - 30°C$ at $I_{\text{inj}} = 80$ mA. The top curve represents the output at 10°C, the bottom curve

4. Experimental Work

shows output at 30°C; the curves in between correspond to temperatures of $11 - 29$°C, respectively. The data plotted for $I_{inj} = 80$ mA also exists for $I_{inj} = 100$ mA, 120 mA and 140 mA. All 4 datasets show the same behavior as in fig. 4.5: an increase in SOA temperature results in an increase of the center wavelength of the gain curve and an increase of FWHM.

For quantification of the effect, all curves are approximated by the Gaussian amplitude function, given in eq. 4.1. The center wavelength of the gain curve and its FWHM are calculated from the resulting functions. This data is plotted in fig. 4.6 a and fig. 4.6 b, respectively.

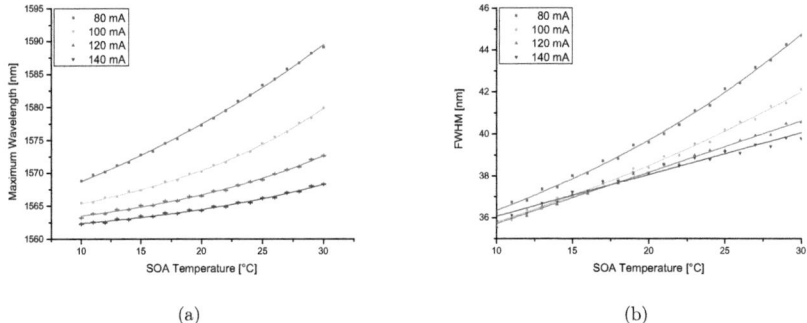

(a) (b)

Figure 4.6.: (a) Measurements on the center wavelength of the gain curve of the SOA at $I_{inj} = 80, 100, 120$ and 140 mA for SOA temperatures from $\Theta_{SOA} = 10 - 30$°C.
(b) Measurements on the FWHM of the gain curve of the SOA at $I_{inj} = 80, 100, 120$ and 140 mA for SOA temperatures from $\Theta_{SOA} = 10 - 30$°C.

The data shown in fig. 4.6 a exhibits an exponential dependence of center wavelength of the gain curve on temperature, which holds for an approximation of the first order exponential function, eq. 4.2.

$$y = y_0 + f_1 \cdot \exp^{x-x_c/f_3}, \qquad (4.2)$$

where x and y are the independent and dependent variable, respectively. y_0, x_c and f_1, f_2 are free fitting parameters. Evaluation of the first derivative of eq. 4.2 at 15°C delivers an estimate for the tuning response: 0.87, 0.49, 0.31 and 0.21 nm/°C for 80, 100, 120 and 140 mA, respectively.

The change of gain curve FWHM is evaluated in the same manner as the shift of gain curve center wavelength. After approximating all curves by eq. 4.1, FWHM of this fit is plotted

4.1. Characterization of the Enhanced Laser System

against SOA temperature for each injection current setting. The results are shown in fig. 4.6 b. The expected trend of an increase of FWHM with increasing temperature can be observed for all four injection current settings. As in fig. 4.6 a, the data follows an exponential dependency quite well. Hence, an exponential fit of the form shown in eq. 4.2 is applied. The evaluation of its first derivative at 15°C delivers an estimate for thermal tuning response in this temperature range: 0.33, 0.27, 0.24 and 0.20 nm/°C for 80, 100, 120 and 140 mA, respectively.

4.1.3. Methods of Manipulating Threshold Current

As mentioned in sec. 3.2.1 and sec. 2.1.5, threshold current is dependent on cavity loss per round trip and can be tailored by means of this parameter. More loss means an increase in threshold current.

In order to show how exactly our system responds to introduction of extra loss in terms of a change in threshold current, firstly a detailed look at the variable coupler V_i has to be taken and its role for loss introduction explained. Secondly, the effect of this loss on threshold current needs to be measured and analyzed. Finally, a direct correlation for threshold current behavior as a function of coupler manipulation is shown.

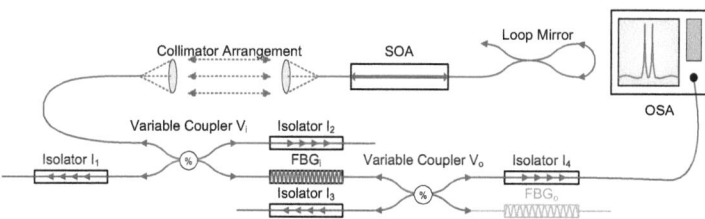

Figure 4.7.: Schematics of the setup used for the measurements of threshold current on mode M_i.

Cavity loss from V_i

Figure 4.8 a shows the behavior of the coupler V_i in the configuration implemented into the setup. For low settings on the micrometer knob of the coupler coupling ratio changes only insignificantly; for values above approx. 30 a.u. a more pronounced change in coupling ratio is observed. As a component of the final setup the loss introduced by coupler V_i doubles in

4. Experimental Work

significance compared to the single pass measurement presented in fig. 4.8 a. This is due to the fact that light passes it once in each direction per round trip. Mathematically it means that the normalized loss L_{V_i} to the laser system, which is introduced by coupler V_i, equals

$$L_{V_i} = 1 - (T_{\text{in}\to\text{tp}} \cdot T_{\text{in}\leftarrow\text{tp}}), \qquad (4.3)$$

where $T_{\text{in}\to\text{tp}}$ represents the transmission from the in to the tp-port of V_i and $T_{\text{in}\leftarrow\text{tp}}$ the transmission from the tp- to the in-port. This calculation of loss from coupler setting is depicted in fig. 4.8 b and bases on the broad band transmission measurements described in sec. 3.2.1.

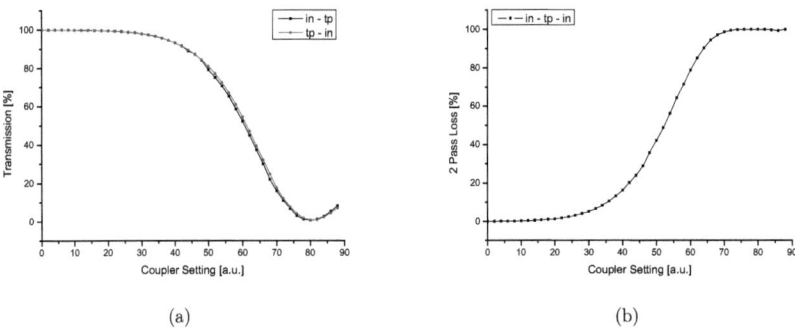

Figure 4.8.: (a) MEasured coupling ratio for coupler V_i between in- and tp-port.
(b) Calculated loss for 2 passes (in-tp, tp-in) on coupler V_i.

Threshold Current from Cavity Loss

This section deals with how threshold current is affected by the loss L_{V_i} introduced by coupler V_i. All threshold measurements are done on mode M_i at 1542.0 nm. The optical output is taken at V_i behind Isolator I_2. The detailed configuration of the setup is depicted in fig. 4.7. For the measurements the SOA is set to a defined temperature of 25°C and V_i adjusted to fixed coupler settings of 0 to 90 a.u. in increments of 2 a.u.. Coupler V_o is set to couple all light out of the setup, completely suppressing oscillation on M_o at 1542.7 nm. FBG_o is hence shown in gray. For each coupler setting of V_i, injection current is scanned from 85 to 120 mA in increments of 1 mA, recording spectra averaged over 20 readings for each combination of coupler setting and injection current. The spectra are evaluated for each coupler setting identically as in sec. 3.2.1, delivering threshold current for each setting of V_i.

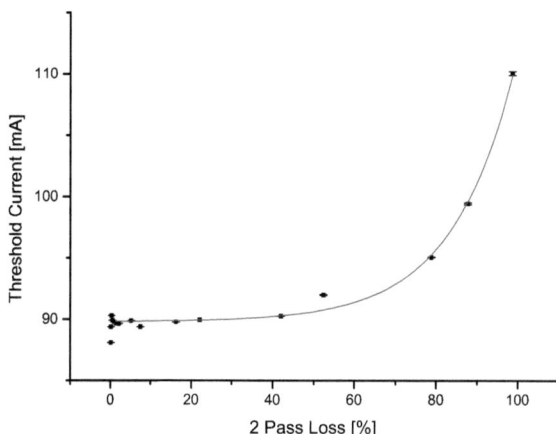

Figure 4.9.: Exponential approximation of threshold current as a function of the 2 pass loss shown in fig. 4.8 b.

Figure 4.9 shows threshold current as a function of loss introduced into the laser resonator by coupler V_i. The rather insignificantly small error bars result from the quality of the linear regressions applied to an intensity vs. current dependency in an intermediate step. Their transfer into threshold current is calculated by Gaussian error propagation. They do not account for systematic errors of the measurement, which shows especially in the low loss region. The dependency of threshold current on loss shows to be exponential, which is emphasized by the first order exponential fit function, eq. 4.2, displayed along with the measurement data in fig. 4.9.

4.1.4. Spectral Characterization of the Laser Modes

Mode FWHM

For ICAS to fully profit from mode competition effects laser line width must be smaller than absorption line width, as stated in sec. 2.3. Hence, knowledge of the laser system's line width is required to obtain reliable information from its response to selective absorption inside the

4. Experimental Work

resonator. This section deals with characterization of the FWHM of the laser system developed in this work. FWHM dependence on injection current and SOA temperature is shown and the whole system's improvement over the system described in [48] demonstrated.

In the acquisition of the data used for this characterization, the setup in fig. 4.1 is used with the OSA connected behind Isolator I_1, as the spectra obtained here are free of ASE. Firstly, SOA temperature is set to the defined values. After temperature stabilization, injection current of the system is increased from 55 to 150 mA for $\Theta_{SOA} = 15°C$ and from 70 to 150 mA for $\Theta_{SOA} = 20$ and 25°C. Averaged over 20 individual readings, the spectra obtained have a form similar to the red curve in fig. 4.2 b. Each spectrum is manually analyzed for FWHM. The data taken at an SOA temperature of 15°C is shown in fig. 4.10. The behavior of FWHM is discussed basing on this data only, as all three data sets share general characteristics.

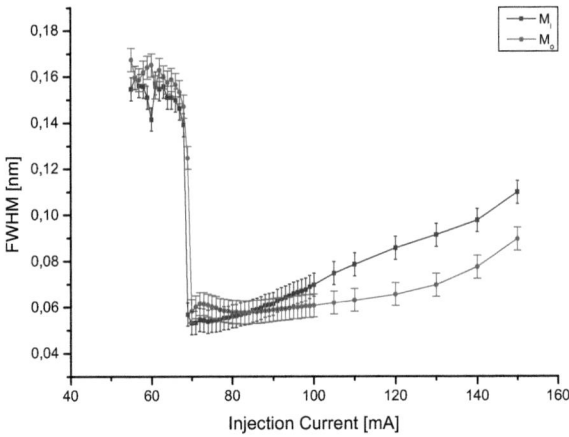

Figure 4.10.: Measured FWHM of M_i and M_o at SOA temperature of $\Theta_{SOA} = 15°C$. The drastic drop at 65 mA corresponds to laser threshold.

The FWHM at low currents corresponds to the reflection bandwidth of the FBG, which is specified at 168 pm by the vendor. At threshold a drastic drop of FWHM is observed, which occurs at the same injection current for both modes. The laser lines that develop exhibit an FWHM of <0.06 nm, which is significantly narrower than aforementioned reflection bandwidth. In the plots . For the curves recorded at 15 and 20°C FWHM remains at <0.06 nm directly above threshold and then gradually increases with increasing injection current. The most

4.1. Characterization of the Enhanced Laser System

important result is that for all three curves the FWHM of the lasing modes is on the order of magnitude of 0.06 nm, directly above threshold which is an improvement over the system described in [89].

Mode SNR

The sensor system bases on the evaluation of suppression of a laser mode by selective loss in the resonator. Evaluation of SNR, mode intensity over the noise level, yields important information on intensity response of a laser mode to such absorption.

For conventional laser diodes, SNR increases gradually with increasing injection current until the fundamental mode of the laser reaches lasing threshold. Then further increase of I_{inj} mainly goes into the main mode, whose intensity drastically increases over any other modes of the laser. This is reflected in SNR as a drastic increase. Finally, SNR stabilizes and subsequently decreases slightly, when the gain medium of the laser becomes saturated on the main mode and other modes gain in intensity.

Here SNR is investigated based on the data obtained from the measurements performed in sec. 4.1.4. All spectra recorded are analyzed for noise intensity $\mathscr{I}_{\text{Noise}}$ and mode maximum intensity $\mathscr{I}_{\text{max}}$. With a noise level of -91 dBm, SNR is calculated by eq. 4.4

$$\text{SNR} = |\mathscr{I}_{\text{max}} - \mathscr{I}_{\text{Noise}}| \tag{4.4}$$

Figure 4.11 shows SNR for M_o (dashed) and M_i (solid) versus injection current for SOA temperature of 15, 20 and 25°C.

Mainly the increase of threshold current with temperature is observed. The relation between SNR of M_i and M_o becomes apparent: the SNR of M_i increases rapidly until it reaches a saturation intensity, at which it shows a very rapid transition into a stable value of about 55 dB followed by a small decrease. M_o shows a much more gradual behavior: initially it does not increase as steeply as M_i, but then also does not flatten as quickly, giving way to a steady increase surpassing the SNR value of M_i, reaching values of up to 60 dB at high currents. The reson for this behaviour is not completely clarified. As the laser gain provided by an SOA is not strictly homogenously broadened, it is assumed that the different ensembles providing laser gain to the modes respond differently to the increase of injection current. This might result in not completely equal behaviour of SNR for the two modes at high current. For a definite answer, however, a detailed investigation is necessary, which lies outside of the scope of this

4. Experimental Work

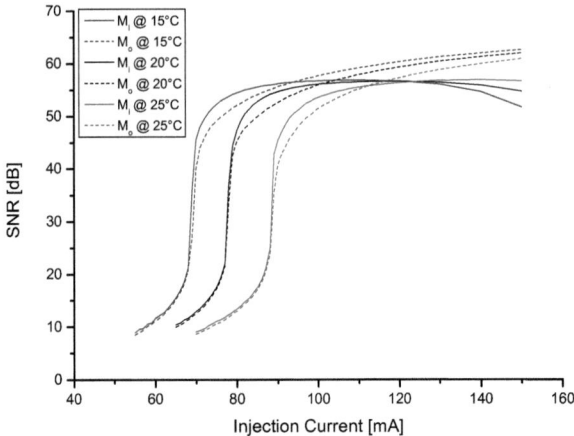

Figure 4.11.: SNR measured as a function of injection current for M_i and M_o at SOA temperatures of 15, 20 and 25°C.

work.

4.1.5. Wavelength Tuning Response

Mode Tunability

As a result of the introduction of variable couplers, described in sec. 3.2.1, the characteristics of the optical output of the laser system under thermal tuning of the FBG has changed significantly. This section deals with these changes and demonstrates the improvements under the perspective of using the system as a sensor device.

Laser output under tuning for the system created in [48] is depicted in fig. 3.3 b in sec. 3.1.3. Here, in fig. 4.12, the spectrum of the tuning measurement at 33°C is extracted and displayed in comparison to the same situation for the system improved in this work.

For the system from [48] even the initial state at 33 °C does not show equal mode intensities: $\mathscr{I}_i = -21$ dB at 1542.01 nm and $\mathscr{I}_i = -17.4$ dB at 1542.50 nm - a difference of 3.6 dB, ($> 50\%$) to begin with. Under tuning to longer wavelengths M_o furtherly increases in magnitude over

4.1. Characterization of the Enhanced Laser System

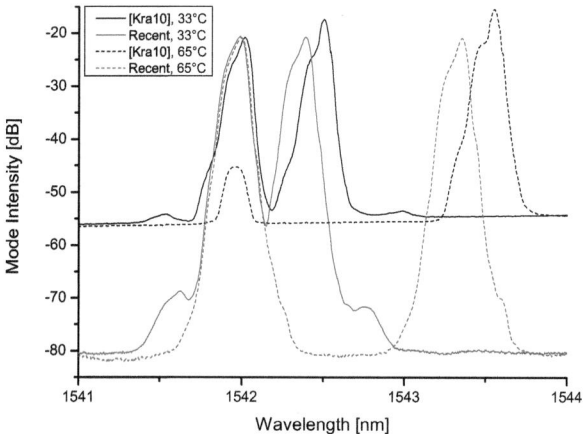

Figure 4.12.: Comparative measurement of the output spectra of the laser system from [48] and the one developed in this work at $\Theta_o = 33$ and 65°C.

M_i. This is due to the fact that it moves up the slope of the SOA's gain curve towards its center, where it receives significantly more gain then M_i. In the situation of maximal mode spacing, denoted by the dashed lines in fig. 4.12, M_i is even fully supressed as a result of mode competition for the system by [48]. The resulting difference in mode intensity is as large as 29.9 dB and very far away from the equlibrium, which is needed for highly sensitive response of the laser system to selective absorption inside the resonator.

With the variable couplers, introduced in sec. 3.2.1, a means of equalizing mode intensity at any given wavelength position or spacing is implemented into the laser system. To show their effect on laser output under mode tuning the measurement leading to fig. 3.3 b is repeated with the improved system developed in this work.

Measurements are taken at SOA temperature of 15°C and injection current of 125 mA. The FBG are set to the same initial temperatures of 24°C and 33°C for FBG_i and FBG_o, respectively. To tune M_o, the temperature of FBG_o is increased to 65°C and subsequently cooled back down to 33°C in increments of 1°C. After each heating or cooling step, V_o is used to re-equalize mode intensities. Then spectra, averaged over 20 measurements to minimize noise, are saved.

While the output of the initial and final state of the new system is already included in fig. 4.12,

4. Experimental Work

the full set of data, corresponding to fig. 3.3 b is displayed in fig. 4.13 a. Figure 3.3 b is shown alongside to ease comparison of the results to the system by [48].

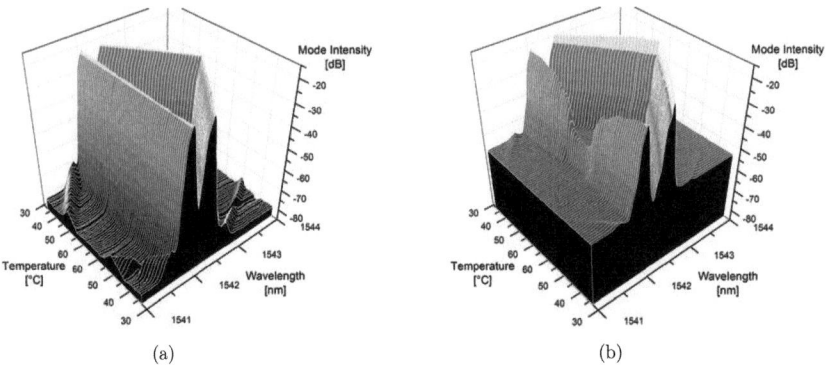

(a)　　　　　　　　　　　　　　　　(b)

Figure 4.13.: Full data set of the thermal tuning response measurement
(a) on the laser system developed in this work
(b) on the laser system from [48].

The improvement is very obvious. Under tuning of M_o, the full intensity of M_i can be maintained throughout the entire measurement, regardless of the position of M_o. Although mode competition does still take place, the laser modes are kept at equilibrium by introduction of an adequate amount of loss by V_o. The difference in mode intensities $\Delta \mathscr{I} = \mathscr{I}_o - \mathscr{I}_i$ under tuning for both systems is shown in fig. 4.14 b. In the recent system $\Delta \mathscr{I}$ stays < 0.15 dB throughout the whole tuning range, whereas the breakdown of M_i in the system developed by [48] expresses itself by a very large $\Delta \mathscr{I}$ of up to 29.9 dB at FBG temperature of 65°C.

The tuning response to FBG heating is shown for both systems in fig. 4.14 a. As the tuning mechanism described in sec. 3.1.3 has not been changed, the slope indicating the tuning response of FBG_o is very similar for both systems. The offset between the two is a result of a relaxation effect of the glue used to attach the FBG to the heated Aluminum blocks. When heating the blocks, the glue holding the FBG on the blocks is also subject to heating. It is assumed that although specifications state otherwise, it does not stay mechanically fully stable under heating, which results in allowing slight relaxation of the fiber on each heating cycle. Repeated heating and cooling cycles in between the measurements taken for [48] in late 2008 and the recent system in early 2010 have hence left their trace in the offset of wavelengths addressable by thermal tuning. The parameters of the linear regressions applied to the data in fig. 4.14 a is summed up in tab. 4.1.

4.1. Characterization of the Enhanced Laser System

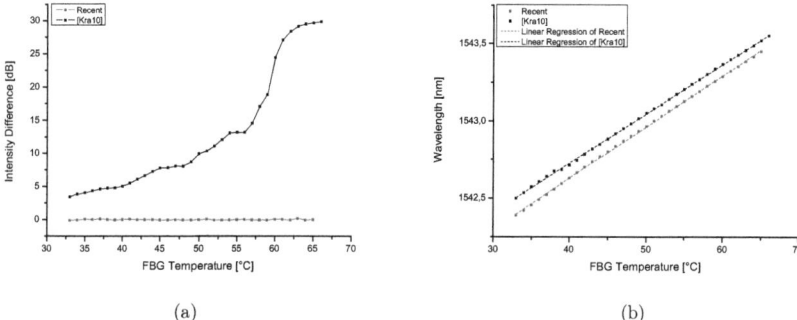

Figure 4.14.: (a) Measured intensity difference $\Delta \mathscr{I} = \mathscr{I}_\mathrm{o} - \mathscr{I}_\mathrm{i}$ under thermal tuning for both laser systems.
(b) Measurement on the tuning response of the laser system by [48] and the one developed in this work.

	System by [48]	Recent Laser System
Intercept (nm)	(1541.452±0.006)	(1541.306±0.006)
Slope (pm/°C)	(31.9±0.1)	(33.1±0.2)

Table 4.1.: Linear regression parameters for the tuning response of the laser system by [48] and the recent one developed in this work.

4. Experimental Work

Mode stability and settling time is unchanged to what is presented in sec. 3.1.3.

As a conclusion to this investigation it can be stated that the tuning mechanism has significantly improved and is capable of reliably establishing a stable intensity equilibrium between M_o and M_i, regardless of any other laser system parameters.

4.1.6. Measurements on Improved Wavelength Stability

In sec. 3.2.3 the problem of thermal instability in the FBG feedback system is discussed as it is implemented and used in [48] and a measurement of its thermal characteristics presented. This section deals with the effect of thermally uncoupling the FBG feedback system from environmental influences. The improved system is characterized and the results presented.

In fig. 3.12 the wavelength stability of the FBG is shown for the system described in [48]. A plot of the same measurement on the FBG system in a Styrofoam enclosure is presented in fig. 4.15. Figure 3.12 from sec. 3.2.3 is shown alongside to ease comparison.

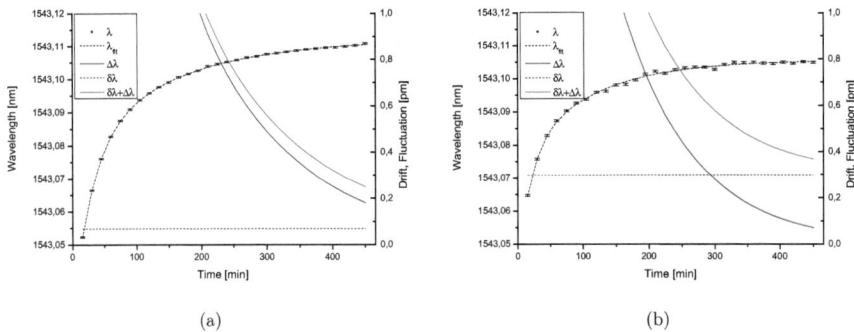

(a)   (b)

Figure 4.15.: (a) Stabilization behavior of the FBG feedback system in the laser system developed in this work, where final stability is determined by long term drift, in comparison to
(b) the laser system from [48], where final stability is determined by thermal fluctuation of the environment.

It is observed that the entire data set in fig. 4.15 a exhibits much smoother characteristics than the one in fig. 4.15 b. The second order exponential fit, eq. 3.2, approximates the data very well, which also shows in the mean magnitude of fluctuations. They are reduced from 0.3 pm to 0.07 pm, a relative change of 76%.

4.1. Characterization of the Enhanced Laser System

The reduced level of environmental fluctuation on the FBG system is additionally reflected in the stabilization time $t_{stab.}$, the time at which the wavelength change transits from being dominated by the thermal inertia of the FBG system to being dominated by environmental fluctuations. As expected, it moves from about 300 min to larger time, even well beyond the total time frame of the measurement. At the end of measurement however, the total fluctuation of wavelength is at 0.3 pm, an improvement over the previous system without thermal insulation of 0.17 pm. Still, stability is ultimately limited by the environmental fluctuations, which range at only <0.07 pm.

As stabilization time is not critical at this point, the increased stability at the expense of stabilization time over the previous system is very welcome, as FBG wavelength can be set much more accurately. Additionally this increased stability enables better comparability of results obtained in long term measurements, as the drift of the system, which is negligible after the system entered the time regime of $t \gg t_{stab.}$, where total fluctuation is dominated by environmental influences.

To furtherly improve overall FBG system performance, thermal design must be refined. Better balance of thermal inertia and heating power, as well as faster heating response and adding the possibility of cooling the FBG, rather than just heating, a functionality provided by i.e. Peltier elements, are factors that vastly influence overall stabilization performance for the better. This issue is considered in the future development of the laser system.

4.1.7. Consequences for the POC

In the previous section the laser system is characterized in respect to its application as a sensing device for ICAS, focusing on three aspects:

1. Characterization of SOA gain curve under variation of injection current and temperature
2. Threshold condition and optical properties of the lasing modes of the system
3. The improved wavelength tuning mechanism

Here the results are summed up and consequences for the most desirable operating conditions of the laser system developed in this work discussed.

4. Experimental Work

Gain Shift

Gain curve blue shift from carrier density increase is measured for temperatures of 15, 20, 25 and 30°C in the injection current range from 5 − 150 mA spanning a wavelength range of 65 nm. Temperature dependent red shift is measured for fixed injection currents of 80, 100, 120 and 140 mA for a temperature range from 5 − 30°C. Below 80 mA it is quite linear, above 80 mA it stabilizes. Temperature redshift follows a weak exponential dependence and linear approximation is still justifiable. For both measurement series also the behavior of gain curve FWHM is analyzed. Under injection current increase FWHM decreases until it reaches a stable value for injection currents above 80 mA. Under increasing of SOA temperature FWHM grows. All results are summed up in tab. 4.2

	Slope	Valid for
Blue Shift	−0.5 nm/mA	all temperatures
Red Shift	+0.87 to +0.21 nm/°C	80 − 140 mA
$FWHM_{Gain}(I_{inj})$	−0.33 to −0.2 nm/mA	80 − 140 mA
$FWHM_{Gain}(\Theta_{SOA})$	stable @ 45 nm/°C	

Table 4.2.: Results of the characterization of gain curve behavior under varied temperature and injection current.

From the perspective of gain curve behavior a few guidelines can be derived as to which parameter ranges of Θ_{SOA} and I_{inj} are most beneficial for the operation of this laser system as a sensor device based on ICAS, which favors stable FWHM, and predictable, favorably linear, tuning characteristics for center wavelength manipulation.

Center wavelength shows linear tuning characteristics from injection current variation in fig. 4.4 a for all four curves until it stabilizes at injection current greater than 80 mA. Center wavelength red shift resulting from temperature change, as shown in fig. 4.6 a, shows no features in preference of any specific temperature or current range.

Gain curve FWHM generally decreases with an increase of injection current until it stabilizes at values above 60 − 80 mA in fig. 4.4 b depending on SOA temperature. The point, where FWHM stabilizes, moves to higher currents with increased temperature. Where the curve taken at $\Theta_{SOA} = 15°C$ begins to stabilize at roughly 60 mA, the curve taken at 30°C still changes noticably. Low temperatures also seem to result in better output characteristics than high ones. This trend is also backed up by fig. 4.6 b.

Comparing the response of FWHM and center wavelength to SOA temperature and injection current variation in the parameter range addressable by our setup, a much stronger response to

4.1. Characterization of the Enhanced Laser System

manipulation by means of injection current is observed. Hence, injection current tuning is more efficient than temperature tuning and has a greater impact on deciding for an optimal operating parameter range. From the perspective of gain curve characteristics it is recommended to bias the system well above threshold and keep SOA temperature low.

Threshold

In sec. 4.1.3 threshold current is found to be equal for M_o and M_i and is reached between 90 and 110 mA depending on coupler V_i. So, what are the findings that are important for the application of this laser system as a sensor system?

This system is based on optical spectroscopy and deals with several parameters, such as selectivity, sensitivity and stability. Its selectivity is dealt with by the FBG. Sensitivity and stability are highly related to gain and loss, which is influenced by the couplers discussed above. For a highly sensitive sensor it is important that resonator conditions are well maintained during sensor operation in order to obtain reliable results. This means that a delicate and quick mechanism is needed to equalize the relative mode intensity prior to measurement by shifting the gain curve via threshold carrier density. It is a very attractive approach compared to the coarse manipulation of coupler V_o (the coupler NOT discussed in this section), as it is much more delicate and can be accomplished solely by variation of electric parameters. However, for this to work, the laser system must be brought into an operating range that will show an effect from this delicate mechanism. This initialization can be achieved by introducing broad band loss into the system through manipulation of the coupler V_i.

From a sensitivity point of view, operation of the laser system in a sensing environment means to operate it in a range of coupler setting that is most responsive. For high sensitivity applications it is hence beneficial to configure a system in a way that coupler V_i is set to 30 to 65 a.u., according to fig. 4.8 b. That this range is still in what the SOA can deal with in terms of injection current is demonstrated by fig. 4.9.

It must be noted, however, that these considerations are rather hypothetical design guidelines for a later evolution of a laser sensor based on this system, which is actually targeted at much smaller absorption than what is necessary for a proof of concept with liquids, the goal of this work. Here, the effect of shifting the gain curve by coupler induced broad band loss is not expected to be very critical; operating range of the coupler has to be considered in a trade of with the results of other laser parameters evaluated in this section.

4. Experimental Work

Laser Modes

M_o and M_i are characterized for FWHM and SNR at $\Theta_{SOA} = 15, 20, 25$ and $30°C$ varying I_{inj} from 60 to 150 mA. Mode FWHM is found to be < 0.06 nm in lasing. Consequently, based on merely the analysis of the FWHM of this setup, the operation of the sensor system slightly above threshold is beneficial for two reasons:

1. FWHM is smallest directly above threshold, which allows very selective addressing of narrow absorption lines with little crosstalk from adjacent ones.

2. Slightly above threshold FWHM remains small in a range of about 10 mA above threshold. This allows for an adjustment of the gain curve to the desired position by means of injection current without much influence on sensor response.

SNR is found to be as high as $55-60$ dB depending on injection current. From the SNR characterization presented above, a current region in which the change in SNR is as meaningful as possible is to be identified. Based on fig. 4.11 it makes sense to distinguish two regimes:

1. The critical regime close to threshold for highly sensitive response

2. The regime well above threshold for gradual but predictable response

Both regimes have an advantage over each other in sensing depending on the specific task.

In a scenario where a specific critical value of absorption (by a poisonous substance i.e.) has to be sensitively detected, operation very close to threshold of the laser device is recommended. In that case the laser system can be tailored in such way that the critical absorption barely suppresses one mode just below threshold, resulting in a very clear intensity drop. As threshold is quite delicately dependent on system loss, the system must be very well designed, installed and stabilized. In that case it could also be used for detecting extremely low absorptions. The mode most suitable for measurements in this configuration is M_i, as its response towards changes in injection current slightly above threshold is more pronounced than for M_o.

In a scenario where absorption is of larger magnitude and different values of absorption are to be detected, like in the POC measurements attempted in this work, operation of the laser system well above threshold is beneficial. In this case too precise tailoring of the device is not necessary and the drastic nonlinear response of M_i at threshold unwanted. For such a situation the sensor should be operated in a way that absorption is more pronounced on M_o, as its intensity change is more pronounced than that of M_i in the regime above threshold. All that needs to be ensured in laser operation is that both modes remain in the lasing state (above

4.1. Characterization of the Enhanced Laser System

threshold) throughout the measurement.

Under tuning with the improved setup mode position responds at ~ 30 pm/°C and mode intensities $\mathscr{I}_o$ and $\mathscr{I}_i$ can be kept in equilibrium at any time by adjusting V_o, which is very beneficial in a sensing application from a sensitivity point of view. Wavelength stability is improved from ultimately 0.3 pm to < 0.07 pm by uncoupling the FBG feedback system from thermal fluctuations of the environment by implementing insulation. From the point of view of wavelength tuning no preference of operating parameters for the POC are derived.

Making the Trade Off

Giving a reccommendation for the most suitable operating parameters is a trade off between different contrary tendencies. Where high injection currents are favorable from the perspective of gain curve behavior, currents close to threshold are beneficial from the point of view of mode FWHM and SNR. Coupler setting and choice of sensing and reference mode depend on the application. The general recommendation for operating parameters in respect to reliability and reproducibility of results in the POC based on the findings in this sec. 4.1 is to use M_o as the sensing mode and M_i as reference mode and to operate the laser system at low SOA temperatures and rather well above threshold. In these conditions the system is expected to be quite stable at the expense of sensitivity, but still very capable of quantitative measurements on the very coarse absorption peak of Propofol. In a future product, the more sensitive regime close to threshold can be targeted.

4. Experimental Work

4.2. Chemical and Optical Characterization of Propofol and Solvents

This section deals with the chemical and optical properties of the target substance for the POC measurements, Propofol, and presents different solvents in which the POC is attempted. Firstly, Propofol is introduced and measurements on its optical properties presented, which are important for system design, conduction of the POC measurements and evaluation of the results. Secondly, the optical characterization of four different substances that are used as solvents in the measurements on varied Propofol concentration is presented.

4.2.1. What is Propofol?

Propofol is an intravenous anesthetic agent used for induction and maintenance of anesthesia as well as sedation [12] due to its rapid effect, good recovery characteristics and small side effects compared to other drugs used in the field. It is commonly applied in the form of an emulsion of water and Soy Oil. It was introduced into the market under the name of Diprivan in the US in 1986, and in Europe in 1996. To date it is well established and available as a brand- and generic drug.

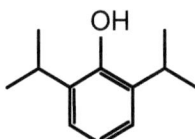

Figure 4.16.: Structure of Propofol.

Propofol Data	
Molar Mass	178.27 g/mol
Density	0.96 g/cm^3
Melting Point	18°C
Boiling Point	256°C
Vapor Pressure	0.4 Pa
Solubility in H_2O	124 mg/l
Refr. Index $n20/D_{Propofol}$	1.514

Table 4.3.: Properties of Propofol [85].

Figure 4.16 shows the chemical structure diagram of Propofol. From a chemical point of view it is a phenol derivative. Its IUPAC name is 2,6-di(propan-2-yl)phenol and its structure formula is $C_{12}H_{18}O$. Table 4.3 lists its most important data.

Propofol reacts with atmospheric oxygen and requires to be stored under nitrogen atmosphere. It is flammable and its fumes are toxic and explosive [88]. Its low solubility in H_2O (tab. 4.3) results from its rather unipolar nature, as its only OH group is shielded. It is well soluble in

4.2. Chemical and Optical Characterization of Propofol and Solvents

Hexane and Methanol and hardly in water [66]. For the experiments conducted in this work it is dissolved in 2-Propanol, Acetone, DiChloromethane (DCM) and Soy Oil, which is used as the agent in anesthesia. Table 4.4 lists these solvents above with their structural diagram, refractive index and direct current dielectric constant as a measure for polarity ans solubility [65]. The dielectric constant ϵ of all solvents lies between the values of Hexane ($\epsilon = 1.88$) and Methanol ($\epsilon = 32.70$). As a result they should be good candidates from a solubility perspective.

From an optical point of view a possible solvent is better suited the more similar its refractive index is to that of the liquids cell material. The reason for this is the reduction of the interference effects mentioned in sec. 3.3.2. This argument clearly favors the Soy Oil as a solvent, as it has the highest refractive index of all solvents considered.

Figure 4.17 a [48] shows the absorption coefficient of Propofol versus wavelength, measured by an NIR spectrophotometer[1]. In the wavelength range depicted several very pronounced features can be seen, such as two major peaks in absorption coefficient at around 1420 and 1700 nm. These, however, largely interfere with water absorption. The smaller peak around 1554 nm is characteristic for Propofol and located in a wavelength range in which water absorption is insignificant and which is well in reach of commercial components targeted at telecommunications. Hence, the POC measurements presented in this work are done on this characteristic absorption band. More information can be found in [48].

In order to use this absorption peak of Propofol as a reference for the POC measurements, very precise knowledge in terms of an accurate absolute value for the absorption coefficient of Propofol in this wavelength range is needed. As the possibility of calibrating of the Spectrometer at the University of Kassel are limited, it is not suitable for the exact determination of absolute values of the absorption coefficient in this state. Justified by economic consideration the absorption coefficient is measured by a third party, the PTB in Braunschweig, a well renowned institute of standardization in Germany. The measurements obtained from them are shown in fig. 4.17 b

The data presented in fig. 4.17 b is determined by a transmission experiment using two round absorption cells of 10 and 15 mm in diameter and calculated by means of the Lambert-Beer law, eq. 2.40

The absorption maximum is found at 1545.7 nm and amounts to $\alpha = 0.14$ mm^{-1}. The relative error of the measurement is 15%, indicated by error bars on each data point [78].

---
[1] Perkin Elmer Lambda 900

4. Experimental Work

Substance	Structure	ϵ_r @20°C	n_{eff} @20°C
Propofol $C_{12}H_{18}O$		--	1.514
Hexane C_6H_{14}		1.88	1.375
Methanol CH_4O		32.70	1.328
Vegetable Oil	--	3	1.47
DCM CH_2Cl_2		8.93	1.424
2-propanol C_3H_8O		19.92	1.377
Acetone $OC(CH_3)_2$		20.70	1.359
Water H_2O		80.36	1.333

Table 4.4.: Characteristic values for Propofol [85], Vegetable Oil [74, 90, 61] and solvents [27, 68].

4.2. Chemical and Optical Characterization of Propofol and Solvents

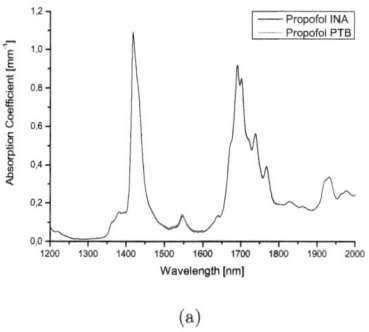 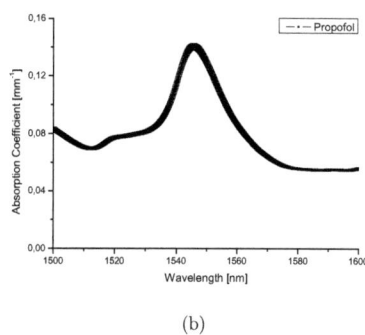

(a) (b)

Figure 4.17.: (a) Broad band absorption coefficient calculated from own transmission experiments on Propofol.
(b) Calibrated absorption coefficient provided by the PTB Braunschweig between 1500 and 1600 nm.

4.2.2. Evaluation of Different Solvents for Compatibility with Propofol

After definition of the wavelength range of interest for the POC measurements, the solvents introduced in sec. 4.2 have to be analyzed for suitability from chemical and optical point of view. A good overview oon the most important parameters is already given in tab. 4.4, where the dielectric constant and the refractive index of the substances are emphasized as a characteristic parameter to determine suitability from chemical and optical point of view. In this section, absorption characteristics are regarded and each solvent evaluated in the scope of these three parameters.

Absorption spectra are obtained from transmission curves recorded by an NIR spectrophotometer[2] to get insight into how the different solvents suggested before absorb in the wavelength range relevant for the POC. The spectrometer used is a two beam system, schematically shown in fig. 4.18.

The initial monochromatic light beam is split into two beams of equal intensity, the reference beam and the sample beam. To initialize a measurement, two identical samples of the substance to be characterized are introduced into the beams and a so called base line is recorded. This serves as a reference for the actual measurements later. After recording the baseline, the sample in the sample beam is replaced by one with increased optical path length, resulting

[2]Perkin Elmer Lambda 900

4. Experimental Work

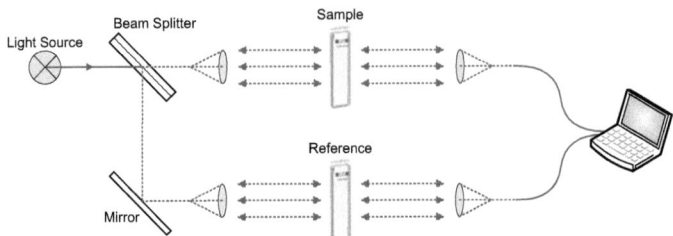

Figure 4.18.: Schematic setup of Perkin Elmer Lambda 900 spectrophotometer.

in reduced transmission over the sample in the reference beam. Both transmission curves are recorded in parallel and subtraction of the two spectra delivers the transmission characteristics of the sample in the sample beam. Monitoring of both beams during this measurement is used to eliminate intensity fluctuations of the light source. The baseline for each one of the solvents characterized is recorded using the 1 mm absorption cells described in sec. 3.3.2. The actual absorption measurements are done by using a 2 and 10 mm cell, resulting in an optical path difference of 1 and 9 mm, respectively. The absorption coefficient is finally calculated by Lambert-Beer law for each of the two curves.

Figure 4.19 a shows the resulting absorption coefficient versus wavelength in the range of 1400 to 1600 nm increments of 0.5 nm. Error bars are omitted here to be able to distinguish the curves clearly. The curves plotted here are the average of the two curves recorded in the 2 and the 10 mm cell for each substance. Figure 4.19 b shows the absorption measurements on the four solvents in the wavelength range around 1545 nm, the range in which our laser system can be operated. For comparison also the absorption curve of Propofol is shown.

All four solvents show flat absorption coefficients with wavelength in the range regarded. The absorption of 2-Propanol is the highest, with $\alpha > 0.35$ mm^{-1}. Soy Oil, Acetone and DCM show much lower absorption coefficients of $\alpha < 0.025$, 0.015 and 0.005 mm^{-1}, respectively. Propofol ranges in between at $\alpha > 0.11$ mm^{-1}. The POC relies mostly on the difference in absorption coefficient sampled by the two modes of the laser system. Hence, the absolute absorption of the solvent is not the most important aspect in terms of general suitability for the POC. Its flatness plays a much larger role. From this point of view all solvents measured seem suitable for the POC.

Absolute absorption of the solvent, however, is not completely negligible, especially in the evaluation of the sensor signal in the concentration measurements. This is discussed in the

4.2. Chemical and Optical Characterization of Propofol and Solvents

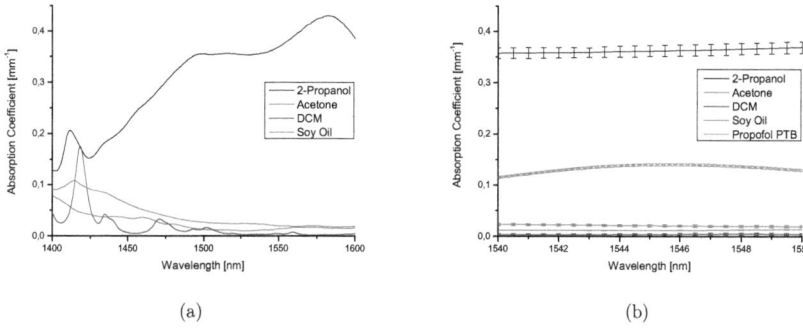

Figure 4.19.: (a) Broad band absorption coefficient calculated from own transmission experiments on the solvents considered for the POC.
(b) Detail of the absorption coefficient of the solvents considered for the POC along with the calibrated absorption coefficient of Propofol provided by the PTB Braunschweig.

following sec. 4.2.3.

4.2.3. How Different Solvents Bias POC Results

As described before, the sensor signal results from the difference in absorption coefficient of the target substance for the two wavelengths λ_i and λ_o. When performing measurements on pure Propofol this is very trivial, as the absorption of the target substance is undisturbed by any other substance. When analyzing a mixture of two substances, like Propofol in a solvent, this becomes a little more complicated, as the absorption characteristics of the solvent also have to be considered.

Generally the absorption on each of the modes depends on the effective absorption coefficient α_eff of the mixture calculated by eq. 4.5

$$\alpha_\mathrm{eff} = \frac{\xi_\mathrm{Solvent} \cdot \alpha_\mathrm{Solvent} + \xi_\mathrm{Propofol} \cdot \alpha_\mathrm{Propofol}}{\xi_\mathrm{Solvent} + \xi_\mathrm{Propofol}}, \qquad (4.5)$$

where ξ_xx is the number of moles of the solvent or Propofol and α_xx its respective absorption coefficient. Introducing α_eff into Lambert-Beer law from eq. 2.40 results in normalized absorption A for each mode. The absorption difference ΔA responsible for a suppression of M_o

4. Experimental Work

is then given by eq. 4.6

$$\Delta A = A_\mathrm{o} - A_\mathrm{i}, \quad (4.6)$$

where A_0 and A_i is the normalized absorption A for mode M_o and M_i, respectively.

For a given concentration of Propofol in a solvent, it becomes apparent in eq. 4.5 that the smaller the absolute absorption coefficient of the solvent compared to the one of Propofol, the larger the impact of Propofol on the absorption coefficient of the solution. In the case of a solution of Propofol in 2-Propanol increasing the concentration of Propofol means two things:

1. The characteristic absorption of Propofol becomes more pronounced in the solution.
2. The absorption effects of 2-Propanol become reduced.

As 2-Propanol inhibits stronger absorption features than Propofol, the resulting absorption spectra of the solution are more biased by the reduction of 2-Propanol related absorption than the increase of Propofol related absorption.

DCM on the contrary, has much weaker absorption features than Propofol. In the case of increasing the concentration of Propofol in DCM, the resulting spectra are expected to reflect more of the increase of the absorption features of Propofol, than the effects of the reduction of DCM.

To put it in a nutshell: the POC measurements are expected to be most straightforward with DCM, Acetone and Soy Oil. Whereas DCM and Acetone have the lowest absorption, Soy Oil is most similar to Propofol in terms of refractive index. All three substances seem well suited for the POC. The measurements performed in 2-Propanol are expected to turn up a bit biased by 2-Propanol features.

5. Proof of Concept

This chapter documents the core achievement of this work. Based on the preceding system characterization, it presents the actual POC measurements, which show the validity of the ideas expressed and recognized in patent DE 10 2004 037 519 B4. It begins by explaining the procedure of data acquisition and giving an overview over the measurements conducted. Subsequently, different methods for data evaluation are discussed and presented. It concludes by impressively demonstrating the laser systems potential as sensor device.

5.1. Data Acquisition

5.1.1. System Initialization

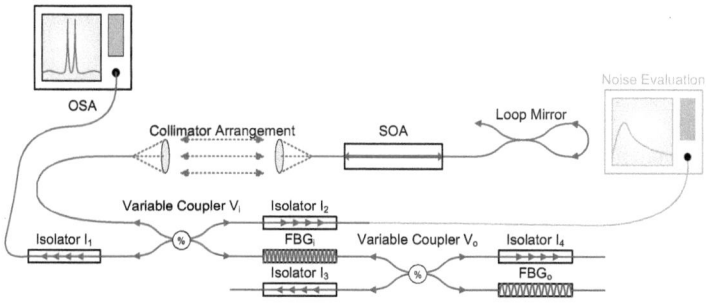

Figure 5.1.: Schematics of the setup used for the POC measurements.

Figure 5.1[1] shows the laser system's configuration during POC measurements. The OSA[2] is connected to the laser system behind Isolator I_1, as clear optical spectra are obtained here

---
[1]also available as a foldout for further reference in the Appendix.
[2]Yokogawa AQ6370

5. Proof of Concept

due to the non presence of ASE, which contributes to noise. Noise is analyzed in parallel; it is recorded behind Isolator I_2 by a fast broad band photo detector[3], amplified by a setup developed and described in [82] and evaluated by an electronic spectrum analyzer[4]. Inside the collimator arrangement one holder is used for the sample cells, the other holds an AR coated piece of fused silica glass for beam offset compensation.

Prior to the POC measurements, the FBG center wavelengths are tuned to the position desired: the SOA is biased above laser threshold and the resulting output spectra are observed in real time, while heating the FBG according to fig. 4.13. When the output wavelengths desired are reached, the SOA is switched off and the FBG heating system is left to stabilize over night before beginning measurements on the following day. The center wavelengths used are 1542.6 nm and 1541.6 nm for FBG_o and FBG_i, respectively, in one measurement 1542.0 nm and 1541.6 nm for FBG_o and FBG_i in the other.

5.1.2. Sample and Cell Preparation

The Propofol used for the POC is delivered in a 100 ml bulk container. It is opened under nitrogen atmosphere and its contents split up, sealed and stored in many small containers holding about 2 ml each at 4°C, where Propofol is in solid phase. Soy Oil is delivered in a sealed 50 ml container, in which it is kept at 4°C for about 3 days before the POC measurements. 2-Propanol, Acetone and DCM are readily available in large quantities and high purities at INA. For the measurements presented in this section new 100 ml bottles have been purchased and used.

All substances are taken out of cooling and left to adjust to room temperature several hours before the beginning of the measurements.

The sample cell used is thoroughly rinsed with 2-Propanol and dried with gaseous nitrogen multiple times before each measurement to remove any remnants of the substances it previously contained.

5.1.3. Taking Measurements

With the FBG heating system stable and the substances adjusted to room temperature the actual measurements are carried out. This involves the following steps to initialize each mea-

---

[3] Newport D-15ir Detector
[4] Rohde & Schwarz FSP30

5.1. Data Acquisition

surement series:

1. The sample cell is thoroughly cleaned as described above, introduced into the collimator arrangement at an angle of 1° towards normal incidence, filled with an initial amount of solvent (2-Propanol, Acetone, DCM or Soy Oil) and closed to avoid evaporation.
2. Current is applied to the SOA and the output spectra observed on the OSA.
3. Injection current is increased until M_i is well above threshold, in compliance with the conclusions of sec. 4.1.7.
4. The total intensity is observed for its maximum, while varying the angle of the beam offset compensation. When the maximum intensity is reached the beam offset is considered compensated.
5. Mode M_o is equalized to M_i by variation of coupler V_o.
6. The initial laser output spectrum in this equalized state is saved as a starting point.

Now the system is ready for the stepwise increase of Propofol. This is done as follows:

1. The Laser beam is blocked and the lid of the sample cell removed.
2. A defined amount of Propofol is introduced into the sample cell.
3. To ensure a homogeneous solution in the cell the entire amount of liquid now present in the cell is picked up and re-introduced multiple times with a very small aperture pipette.
4. The Laser beam is unblocked again.
5. The resulting spectrum is saved.

Steps 1 through 6 are repeated for each increase of Propofol concentration until the output spectra do not show any significant change anymore. The result of this measurement are multiple output spectra of the system, which then have to be correlated with the characteristics of Propofol and evaluated accordingly.

5.1.4. Measurement Overview

Table 5.1[5] shows an overview of all measurements attempted for POC. The index on the left is introduced for simplification of further reference.

---
[5] also available as a foldout for further reference in the Appendix.

5. Proof of Concept

Index	Solvent	Volume/Cell Size	SOA Current	Δλ
1 — ▲ —	Acetone	1 ml / 5 mm	125 mA	1 nm
2 --- ▲ ---				0.5 nm
3 — ■ —		2 ml / 10 mm		1 nm
4 --- ■ ---				0.5 nm
5 — ▲ —	DCM	1 ml / 5 mm	125 mA	1 nm
6 --- ▲ ---				0.5 nm
7 — ■ —		2 ml / 10 mm	160 mA	1 nm
8 --- ■ ---				0.5 nm
9 — △ —	Acetone	0.75 ml / 5 mm	110 mA	1 nm
10	Soy Oil	1 ml / 5 mm	125 mA	1 nm
11	2-Propanol	— —	— —	— —

Table 5.1.: Overview of the measurement taken for the POC. The top measurements on Acetone and DCM are best suited in the following argument for comparability reasons.

The measurements on DCM and Acetone in the top section of tab. 5.1 worked as expected from the considerations in sec. 4.2.3. After introduction of the pure solvent, the laser could easily brought into lasing and calibrated into a mode intensity equilibrium. Propofol dilutes well in Acetone and DCM. The resulting spectra are reproducible and stable. The measurements on Propofol in 0.75 ml Acetone, noted as series 9, is a preliminary measurement for which no comparison at other mode spacing exists. Hence, the measurement series 1-8 are considered as best suited for POC.

The measurements 2-Propanol and Soy Oil in the bottom section of tab. 5.1 are omitted in evaluation. They are not suitable for the following reasons:

- 2-Propanol turns out to be initially too absorbent to bring the laser system into lasing even with the smaller 5 mm sample cell and high injection current of up to 180 mA. A 2 mm cell, which is also available, is unsuitable because of very difficult handling. This eliminates any possibility of performing measurements with 2-Propanol as solvent.

- The measurement on Propofol in Soy Oil fails because of incompatibility between Propofol and the oil. Although a solution of Propofol and Soy Oil is commonly used as injection agent in medical applications, it is not suited for our spectroscopic system, as the solution turns out to be very inhomogeneous, when mixing the liquids inside the sample chamber. In its mediacal application this is not a problem; however, the diffusion of Propofol inside the oil causes strong concentration fluctuations within the small cross section of the laser beam penetrating the solution, which results in extremely unstable spectra that cannot be evaluated. Pre-mixing the two liquids outside the cell and then measuring on the pre-mixed solutions is unfortunately also not possible as the respective previous solution cannot be fully removed from the sample cell without taking it out of the system. That in return resets the calibration.

As a result, the POC for the sensor system is done on Propofol solutions of different concentrations in Acetone and DCM.

5.2. Data Evaluation

5.2.1. Unmodified Output Spectra

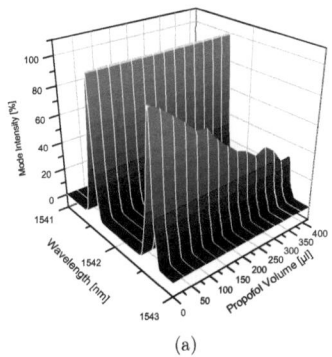
(a)

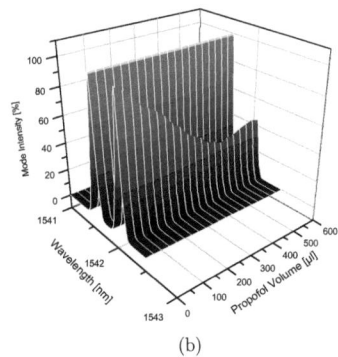
(b)

Figure 5.2.: Normalized output spectra of the laser system with absorption from Propofol
(a) in 1 ml DCM at $\Delta\lambda = 1$ nm (series 5) and
(b) in 1 ml DCM at $\Delta\lambda = 0.5$ nm (series 6).

Generally speaking, all measurements taken and presented in tab. 5.1 show the same behavior, only distinguished by rather small fluctuations. For simplicity the main behavior is discussed based on the two curves presented in fig. 5.2. Where reasonable, other measurements are presented for comparison or to show the big picture.

Figure 5.2 a,b show the output spectra of the system for Propofol in 1 ml DCM at $\Delta\lambda = 1$ and 0.5 nm (series 5 and 6) under increase of Propofol volume in steps of 30 µl. As the spectra are taken behind Isolator I_2 their absolute intensities depend on the setting of coupler V_i. They do

5. Proof of Concept

not reflect the situation inside the resonator at all. This can be understood by the following example: if coupling ratio is increased, more light is coupled out of the system by V_i and the intensity inside the system decreases; however, due to the higher percentage of light coupled out, the intensity that is seen on the OSA increases. The exact calculation of the trade of between the radiation remaining in the laser and the radiation coupled out is not trivial and also bears no information for the evaluation of the sensor response. To circumvent this problem all curves are normalized to the maximum intensity of M_i.

The most obvious observation in fig. 5.2 is the straightforward decrease of M_o over M_i. Where both modes are initially equally intense each addition of Propofol results in further suppression of M_o. These suppression steps also seem equidistant, if the maximum intensity of M_i is taken as a reference point. Additionally, it seems like during the suppression of M_o, its FWHM remains the same. This assumption is dealt with in the following sec. 5.2.2.

At the end of the concentration series the supression of M_o is reduced and a slight increase is observed. This increase is disregarded in the evaluation for the following reason. At the minimum intensity of M_o in fig. 5.2, M_o is already supressed below threshold. Both modes are still affected by a further increase of Propofol in the solution. However, as a result of the resonator effect, this increase of absorption from Propofol now shows a stronger effect on the amplification of M_i, which is still above threshold. As the normalization applied takes the intensity of M_i as a point of reference, its decrease expresses itself in a relative increase of the intensity of M_o.

Figure 5.3 shows a comparison of the spectra of Propofol in 1 ml DCM at $\Delta\lambda = 1$ (series 5) and in 2 ml DCM at $\Delta\lambda = 1$ (series 7). Note that for these measurements not only concentration and cell size are changed, also injection current is increased from 125 mA to 160 mA for Propofol in 2 ml DCM at $\Delta\lambda = 1$ nm (series 7). Despite the changes in three of four parameters, very similar behavior is observed for suppression characteristics.

Two main characteristics can be assumed for further investigation:

1. Mode suppression as a function of Propofol concentration is rather linear.
2. FWHM is unaffected by this suppression.

5.2.2. Comparing Integral Ratio and Maximum Intensity Ratio

The question arising from the observations in the previous section is how to quantify the effect observed most conveniently. In any case the sensor responds to a variation of the concentration

5.2. Data Evaluation

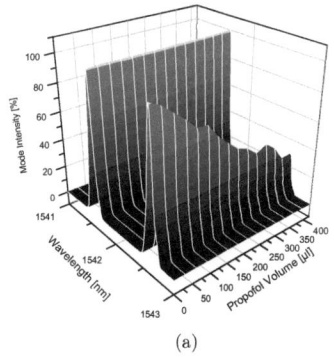
(a)

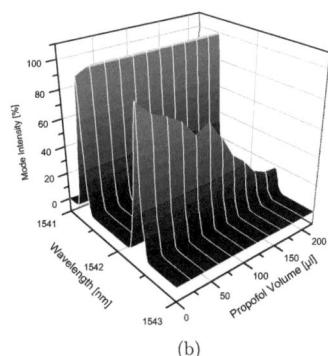
(b)

Figure 5.3.: Comparison of normalized output spectra of the laser sytem with absorption from Propofol
(a) in 1 ml DCM at at $\Delta\lambda = 1$ nm (series 5) and
(b) in 2 ml DCM at at $\Delta\lambda = 1$ nm (series 7).

of Propofol in a solvent with a change in mode intensity. It can be approached in two ways.

The integrated intensity $\mathscr{I}_{\text{int}}$ in a spectral range is given by eq. 5.1, already used in sec. 3.2.1

$$\mathscr{I}_{\text{int}} = \int_{\lambda_1}^{\lambda_2} \mathscr{I}_\lambda(\lambda)\, d\lambda, \tag{5.1}$$

where λ denotes wavelength and $\mathscr{I}_\lambda$ power spectral density, the intensity per single wavelength. The intensity difference of two peaks as shown above can be expressed as their ratio Υ in eq. 5.2

$$\Upsilon_{\text{int}} = \frac{\mathscr{I}_{\text{int}}(M_o)}{\mathscr{I}_{\text{int}}(M_i)}, \tag{5.2}$$

where $\mathscr{I}_{\text{int}}(M_o)$ and $\mathscr{I}_{\text{int}}(M_i)$ are the total intensities of M_o and M_i, respectively.

Another possible indicator for mode intensity is the maximum intensity $\mathscr{I}_{\text{max}}$ of each mode. The intensity ratio then takes the form of eq. 5.3

$$\Upsilon_{\text{max}} = \frac{\mathscr{I}_{\text{max}}(M_o)}{\mathscr{I}_{\text{max}}(M_i)}, \tag{5.3}$$

where $\mathscr{I}_{\text{max}}(M_o)$ and $\mathscr{I}_{\text{max}}(M_i)$ are the maximum intensities of M_o and M_i, respectively.

5. Proof of Concept

In case the FWHM of such peaks is invariant under change of maximum intensity eq. 5.2 and eq. 5.3 should be equal. Hence, a comparison between the two yields information on the invariance of mode FWHM under mode suppression from absorption.

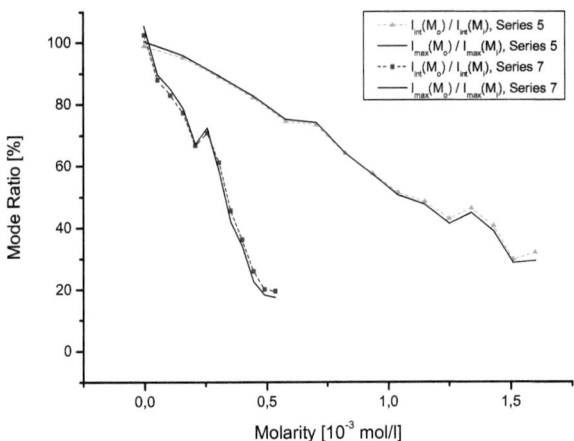

Figure 5.4.: Comparison between Mode Intensity Ratio for $\mathscr{I}_{int}$ and $\mathscr{I}_{max}$ for Propofol in 1 ml DCM at $\Delta\lambda = 1$ nm (series 5) and in 2 ml DCM at $\Delta\lambda = 1$ nm (series 7).

Figure 5.4 shows such a comparison for Propofol in 1 ml DCM at $\Delta\lambda = 1$ (series 5) and in 2 ml DCM at $\Delta\lambda = 1$ (series 7) discussed at the end of sec. 5.2.1. The data is plotted versus the molarity M (mol/l) of the solution, a measure for concentration defined in eq. 5.4[6]

$$M = \frac{\text{moles solute}}{\text{liters solution}}, \qquad (5.4)$$

Although the measurements in fig. 5.4 differ in sample concentration, cell size and injection current, both deliver equal results for the two methods of calculating mode ratio. This verifies the assumption made in sec. 5.2.1 that FWHM of the laser modes is invariant under intensity changes from absorption.

Moreover it qualifies both methods of approaching mode intensity as a well suited means of

---
[6]Alternatively, the molar fraction χ in % or parts per million ppm (mg/kg) can be used. For definitions, see sec. 5.2.3.

5.2. Data Evaluation

quantifying mode intensity variations in the laser sensor system, leaving it up to the application for which to decide.

In any case, from a more practical point of view, the integration in eq. 5.1 seems to be the better approach. Firstly, it accounts for the whole bandwidth of the mode and not only for its very center wavelength. When regarding absorption lines, which are narrower than the FWHM of the lines observed in our system, absorption could happen on the shoulder of one of the modes of this laser system. In that case the method considering the maximum intensity only might fail. Secondly, photo detectors used for measuring the intensity on one of the modes are broad band devices unless equipped with expensive filter technology. Using integration over a broader wavelength range including one of the two individual modes of our laser system is much closer to how a real system works.

For this reason integration defined in eq. 5.2 is used in the further evaluation of the spectra, although the method based on maximum intensity might as well deliver legitimate results.

This section in one sentence:

- Comparison of Υ_{max} and Υ_{int} with concentration shows no significant differences, which means that FWHM is constant under mode suppression from absorption.

5.2.3. Integral Ratio versus Concentration

This section deals with evaluation of mode intensity ratio with Propofol concentration. In order to put the data into perspective, the first data set presented in fig. 5.5 is plotted against three different measures of concentration:

Molarity M from eq. 5.4,

Molar Fraction $\chi = \frac{\text{moles solute}}{\text{moles solution}}$, given in % and

parts per million $= \frac{\text{milligrams solute}}{\text{kilograms solution}}$, given in ppm.

All three measures of concentration are commonly used, depending on the field of work. They differ by a factor dependent on the material constants of the materials involved. As the laser sensor device is capable of being applied to many different measurement tasks, a quick comparison of data in this manner allows easier reference to existing systems.

Figure 5.5 clearly demonstrates the similarity of the data plot under the three definitions of concentration mentioned above. All three graphs show the same curve, only the numbers on

5. Proof of Concept

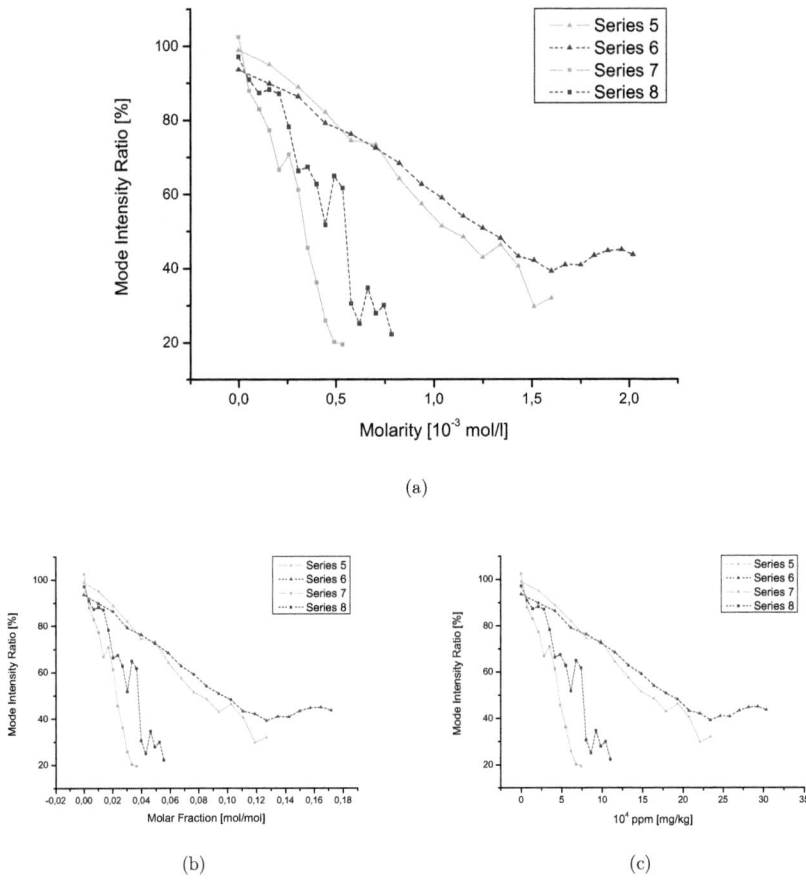

Figure 5.5.: Comparison between different concentration definitions
  (a) Molarity M,
  (b) Molar Fraction χ and
  (c) parts per million 'ppm'.

5.2. Data Evaluation

the x-axis are changed according to the definition of concentration used. For simplicity, Molarity M from eq. 5.4 is used in the further evaluation process.

As hinted at in sec. 5.2.1 a quite linear dependency of Υ on concentration can be observed in fig. 5.5, especially in the domain of 100% > Υ > 70%. The larger the concentration and the smaller Υ becomes, the larger the deviations from the linear behavior. In the domain of 50% > Υ fluctuations dominate and a clear system response to further increase of Propfol concentration cannot be obtained. This behavior also becomes obvious in fig. 5.6, showing Υ versus concentration for all samples measured.

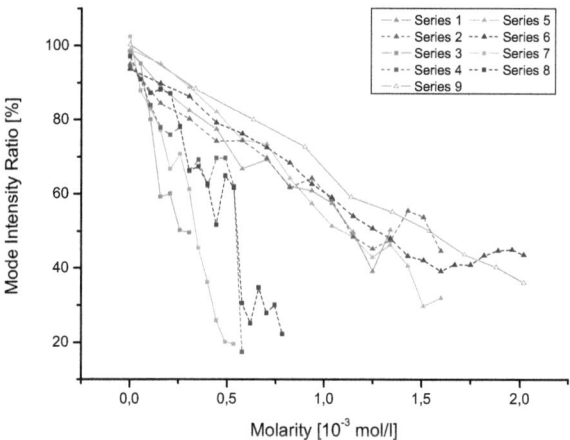

Figure 5.6.: Comparison of Mode Intensity Ratio Υ versus concentration for measurement series 1-9.

The curves measured can be divided into two groups; the ones with triangular symbols, standing for measurements in the smaller 5 mm cell and the ones with square symbols, representing the measurements on the 10 mm cell. A longer optical path through a cell at the same concentration means higher absorption compared to a shorter cell. Hence the curves showing measurements with the 10 mm cell have a bigger slope, which indicates a more dramatic sensor response.

As mentioned before, all curves deviate noticeably from linear behavior, when Υ becomes small. For the measurements taken with the 5 mm cell, this happens when Υ < 50%. For the measurements taken with the 10 mm cell the increase of deviation from the linear case happens

119

5. Proof of Concept

earlier, at $\Upsilon < 70\%$. The curves in the group of square symbols measured at concentrations above 0.0005 mol/l even show a rapid drop almost to $\Upsilon = 0\%$, which corresponds to a complete breakdown of mode M_o. If this breakdown can be exploited for increasing sensor performance in a precisely tailored sensor is an open question whose answer lies beyond the scope of this work and will be subject to future investigation.

To put the findings in this section in a nutshell:

- Sensor response is found to be linear at mode intensity ratio above 70%.
- The slope of the linear response depends on cell thickness: the longer optical path length in longer cells results in more absorption, which expresses itself in greater slopes.
- Sensor response deviates from linear behavior more at higher concentrations.
- For the measurements taken with the 10 mm cell, mode M_o breaks down completely, which might be exploitable to increase sensor performance in future.

5.2.4. Integral Ratio versus Single Pass Absorption

This section analyses the behavior of mode intensity ratio Υ with the absorption difference ΔA from eq. 4.6. Although this seems similar to what is done in sec. 5.2.3, the results provide significantly different information about the laser sensor system, than the behavior of Υ with solute concentration.

The difference between the presentation here and in sec. 5.2.3 consists in the fact that the data in sec. 5.2.3 contains information on data evaluation from the very practical point of view of application, whereas the presentation in this section allows deeper scientific insight into sensitivity resulting from mode interaction. Let us assume that the modes of our laser system are completely independent from each other. This means that mode intensity ratio Υ is only determined by ΔA, regardless of mode spacing. In a plot of Υ versus ΔA all measurements taken should lie on one single curve, regardless of cell size, injection current or concentration, as these parameters are all absent in eq. 4.6.

Very clearly, fig. 5.7 shows a different picture: the measurements do not lie on one single curve, but split up into two individual ones, which each show linear behavior of characteristically different slope. The line of smaller slope is formed only by measurements taken at $\Delta \lambda = 1$ nm, while the line of larger slope is formed by the measurements taken at $\Delta \lambda = 0.5$ nm. Υ responds significantly more sensitive to ΔA for smaller $\Delta \lambda$. The measurement on Propofol in 2 ml

5.2. Data Evaluation

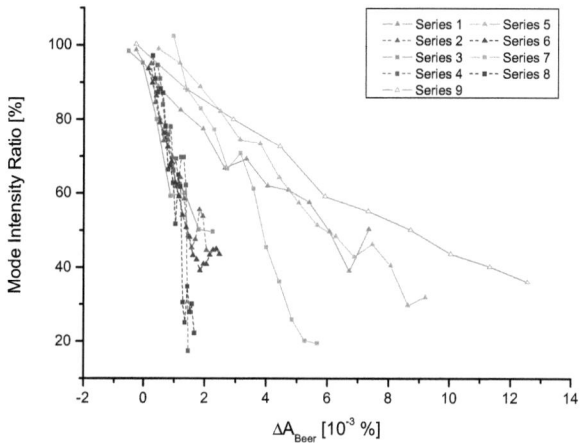

Figure 5.7.: Comparison of Mode Intensity Ratio Υ versus ΔA for measurement series 1-9.

Acetone taken at $\Delta\lambda = 1$ nm (series 3) actually belongs to the line of smaller slope. However, it shows up amongst the measurements taken at $\Delta\lambda = 0.5$ nm. The reason behind this is not clear. Due to fact that, apart from this one measurement, the correlation between slope and $\Delta\lambda$ is otherwise quite straightforward, human error in data handling is assumed to be the problem. Further measurements will clarify the confusion about this.

If the statement that no interaction between the laser's modes means that Υ depends only on ΔA and all data points should lie on one single curve as a result of that, the fact that this is not the case means that there must be an influence on Υ apart from ΔA that has to do with mode spacing. From theory it is known that the smaller FSR between modes, the stronger their competition for gain becomes. Resultingly, the artificially established intensity equilibrium created in the initialization process becomes more sensitive to distortions, such as absorption in the resonator.

The behavior observed here clearly indicates an increase of system sensitivity at small $\Delta\lambda$ as a result of mode competition effects and thereby verify the key measurement principle claimed in patent DE 10 2004 037 519 B4.

For recapitulation:

121

5. Proof of Concept

- Plotting Υ against ΔA results in a straightforward regrouping of curves by mode spacing.
- The slope of curves of these two groups indicates a difference in sensor response with mode spacing resulting from mode competition.

5.2.5. Estimate of Sensitivity Limit

In a final word about sensitivity, a conservative estimate of the sensitivity limit of the system at the operating parameters used in this POC is given. This is done by estimating the minimal sensor response that can be distinguished from noise for the more sensitive configuration at $\Delta\lambda = 0.5$ nm, shown in fig. 5.8 for Propofol in 1 ml DCM at $\Delta\lambda = 1$ nm (series 5) and in 1 ml DCM at $\Delta\lambda = 0.5$ nm (series 6) and calculating the correlated effective absorption coefficient with Lambert-Beer law eq. 2.40.

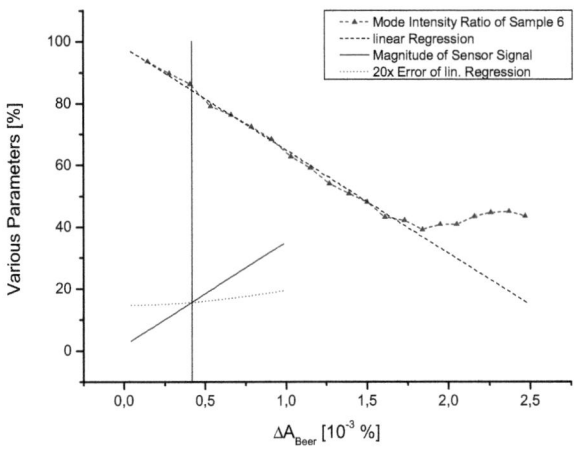

Figure 5.8.: Comparison of Mode Intensity Ratio Υ versus ΔA for measurement series 6.

Figure 5.8 shows the measurements on Propofol in 1 ml DCM at $\Delta\lambda = 0.5$ nm (series 6) along with a linear regression based on the data range from 0 to $2 \cdot 10^{-3}$ %. To estimate the minimal ΔA at which the sensor responds reliably two additional curves are shown in the bottom left.

The solid blue curve is the magnitude of the response of mode intensity ratio $|\Upsilon_{\text{int}}|$. It is defined by eq. 5.5

5.2. Data Evaluation

$$|\Upsilon_{\text{int}}| = |\Upsilon_{\text{int}}(0) - \Upsilon_{\text{int}}(\Delta A)|, \tag{5.5}$$

where $\Upsilon_{\text{int}}(0)$ and $\Upsilon_{\text{int}}(\Delta A)$ are Mode Intensity Ratio for the pure solvent and any respective A, respectively. The dotted red line represents the 20-fold error of the linear regression $\phi_{\Upsilon_{\text{reg}}}(\Delta A)$, calculated by eq. 5.6

$$\phi_{\Upsilon_{\text{reg}}}(\Delta A) = \sqrt{\left(\frac{\partial \Upsilon_{\text{reg}}}{\partial f_2}\phi_{f_2}\right)^2 + \left(\frac{\partial \Upsilon_{\text{reg}}}{\partial f_1}\sigma_{f_1}\right)^2}, \tag{5.6}$$

where f_1 and ϕ_{f_1} are the intercept of the regression with respective error and f_2 and ϕ_{f_2} the slope of the regression with respective error. Both increase with A. Equalization of eq. 5.5 and eq. 5.6 weighted by a factor w, as shown in eq. 5.7, delivers the intersection of the two curves, which determines the absorption difference ΔA, at which the sensor signal Υ_{int} outweighs the statistical error w-fold. Although it seems like a small value of $w = 10$ can be justified based on fig. 5.8, a more conservative value is chosen, as the errors plotted are merely based on the Gaussian error propagation and disregards further systematic errors. For $w = 20$ $\Delta A = 4.18 \cdot 10^{-3}\%$, indicated by the black vertical marker in fig. 5.8.

$$|\Upsilon_{\text{int}}| = w \cdot \phi_{\Upsilon_{\text{reg}}}(\Delta A) \tag{5.7}$$

From these considerations, the minimal absorption difference between the two modes of the laser system detectable amounts to $\Delta A = 4.18 \cdot 10^{-3}$ %. Introducing it into eq. 2.40 and assuming a sample cell length of 5 cm, given by the dimensions of the collimator arrangement described in sec. 3.2.2, delivers the minimal molar absorption coefficient $\alpha_{\text{eff}} = 8.37 \cdot 10^{-5}$ mm^{-1} quantitatively measurable by this system in case one mode is absorption free and the other one fully subject to absorption.

5. Proof of Concept

5.3. POC Summary

At the beginning of this section it is stated that this chapter documents the core achievement of this work: proving the concepts and ideas behind patent DE 10 2004 037 519 B4 valid.

The detailed investigation and careful treatment of samples and solvents in sec. 5.1.2 and the selection of measurement routines for the POC experiments described in sec. 5.1.3 lead to a series of equally interesting and meaningful conclusions about the potential of the laser system developed in this work as a sensor device.

Already from the direct output spectra of the laser system a straightforward response of the system to an increase of Propofol concentration is observed.

Evaluation of data processing options leads to the conclusion that the ratio of mode intensities calculated by integration over each mode, as shown in sec. 5.2.2, is the most suitable depiction of the intensity change from absorption. The behavior of mode intensity ratio Υ_{int} is compared plotted against different common conventions for quantifying concentration that all turn out to be equally suited for analysis of the effects observed. Molarity is preferred for the rest of the analysis, as it is a common convention when dealing with liquid solutions.

As a byproduct of the comparison of different options for data plotting the assumption that mode suppression happens under maintained mode FWHM is confirmed. Plotting Υ_{int} against Molarity yields linear sensor response. Section 5.2.3 shows that its slope is only dependent on cell length regardless of injection current or mode spacing. The latter however plays an important role when plotting Υ_{int} against the difference in Lambert-Beer Absorption ΔA, as explained in detail in sec. 5.2.4. Its behavior clearly proves the existence of mode competition as a sensitivity enhancing effect.

Finally, an estimate of the maximum sensitivity of the laser system as a sensing device is given. It is capable of quantifying absorption coefficients on the order of $\alpha_{\text{eff}} = 8.37 \cdot 10^{-5}$ mm^{-1}.

6. Conclusion and Outlook

Basing on the two mode external cavity semiconductor laser developed in [48], in this work the modifications have been documented, which enhance it to a functional sensing device. The system developed responds quantitatively to the increase of the concentration of Propofol in the solvents DCM and Acetone. The response observed is rather linear and its strength varies with mode spacing. The minimal absorption coefficient detectable, which can be considered the sensitivity limit of the system, is $\alpha_{\text{eff}} = 8.37 \cdot 10^{-5}$ mm^{-1}.

Considering these results, the guiding question of this work, if the concepts and ideas behind patent DE 10 2004 037 519 B4 are valid or not, can definitely be answered positively. They are generally valid and worth pursuing further. There is a quantitative effect observed in the output of the laser system developed, in response to the concentration of Propofol in liquid solutions, with sensitivity depending on mode spacing. In its present state the laser system is capable of detecting differences in absorption signal of $\Delta A = 4.18 \cdot 10^{-3}$%, which is sufficient for measurement on liquid samples. If this is sufficient to quantify Propofol in the gas phase is not clear, as there are no NIR absorption spectra of Propofol available as of now.

How mode spacing quantitatively influences sensor response is still left to be investigated in detail. In this work measurements have been performed at two different mode spacings. The measurements at the smaller of which turned out to react more sensitively. In order to quantify this increase of sensitivity at decreased mode spacing a larger volume of measurements has to be taken repeatedly and a detailed theoretical investigation of the mode competition expected in this system in respect to mode spacing undertaken.

A subject left completely untouched is the short time scale temporal behavior of laser output. With B. Scherer, the author of [76, 77], the existence of mode hops in between the two modes of our laser system on the timescale of 10^{-4} s was discussed. These, however, cannot be detected with the present system, as the integrating time of the OSA[1] used exceeds this short time by far. I recommend the implementation of a chopper to the system. In this case not only mode

---
[1] Agilent 86142B

6. Conclusion and Outlook

hops can be investigated, but also the initial phase in lasing and sensor response in a pulsed operation investigated. It also opens the path to cavity ring down spectroscopy under mode competition, a combination of two very powerful tools.

In order to reach a fully functional sensor system that can reliably assist anesthetists in monitoring and regulating the depth of a patients' sleep during surgery, a few words have to be lost on selectivity, which, as mentioned in the introduction, is also a very important aspect of a sensor concept, when thinking of false alarms. The technology developed in this work is based on the mere absorption of EM-radiation on clearly defined wavelengths. What exactly causes this absorption is not evaluated by the laser in the present state, so response to other substances that absorb on the same wavelengths is possible. This problem, however, only exists in liquid phase, as absorption spectra are broadened to continua here. Measurements in the gas phase have the great advantage of the rather discrete absorption characteristics of gases, which should eliminate most cross talk related issues. In any case the combination of this system with another broad band sensor as proposed in the EU-project 'The Emotion-AAL village'[2] should take care of that problem.

---

[2] www.emotionaal.eu

Bibliography

[1] Poisoning of platinum surfaces by hexamethyldisiloxane (hmds): application to catalytic methane sensors. *Sensors and Actuators B: Chemical*, 40(2-3):117–124, May 1997.

[2] R. Adams, P. Warner, B. Hubbard, and T. Goulding. Decreasing turnaround time between general surgery cases. *JONA*, 34(3):140–148, 2004.

[3] Z. Alferov and R. Kazarinov. Semiconductor laser with electric pumping. 1963.

[4] J. R. Andrews. Low voltage wavelength tuning of an external cavity diode laser using a nematic liquid crystal-containing birefringent filter. *Photonics Technology Letters, IEEE*, 2(5):334–336, August 2002.

[5] P. Atkins. *Atkins' Physical Chemistry*. Oxford University Press, 8th edition, March 2006.

[6] Intraoperative awareness, 2009.

[7] V. M. Baev, J. Eschner, E. Paeth, R. Schüler, and P. E. Toschek. Intra-cavity spectroscopy with diode lasers. *Applied Physics B: Lasers and Optics*, 55(6):463–477, December 1992.

[8] V. M. Baev, T. Latz, and P. E. Toschek. Laser intracavity absorption spectroscopy. *Applied Physics B: Lasers and Optics*, V69(3):171–202, 1999.

[9] V. M. Baev and P. E. Toschek. Sensitivity limits of laser intracavity spectroscopy. In H. Schiff and U. Platt, editors, *Society of Photo-Optical Instrumentation Engineers (SPIE) Conference Series*, volume 1715 of *Society of Photo-Optical Instrumentation Engineers (SPIE) Conference Series*, pages 381–392, February 1993.

[10] X. Baillard, A. Gauguet, S. Bize, P. Lemonde, P. Laurent, A. Clairon, and P. Rosenbusch. Interference-filter-stabilized external-cavity diode lasers. *Optics Communications*, 266(2):609–613, October 2006.

[11] M. Bass. *Handbook of Optics*, volume IV. McGraw-Hill, 2 edition, October 2001.

[12] A. G. B|Braun Melsungen. Propofol-lipuro, October 2005.

Bibliography

[13] A. G. Bell. On the production and reproduction of sound by light. *American Journal of Sciences*, XX(118):305–324, October 1880.

[14] G. Berden and R. Engeln, editors. *Cavity Ring-Down Spectroscopy: Techniques and Applications*. Wiley-Blackwell, November 2009.

[15] J. P. Besson, S. Schilt, E. Rochat, and L. Thévenaz. Ammonia trace measurements at ppb level based on near-ir photoacoustic spectroscopy. *Applied Physics B: Lasers and Optics*, 85(2):323–328, November 2006.

[16] R. Böhm, A. Stephani, V. M. Baev, and P. E. Toschek. Intracavity absorption spectroscopy with a nd3+-doped fiber laser. *Opt. Lett.*, 18(22):1955–1957, November 1993.

[17] R. Böse, J. Miller, and E. Inn. Intensity measurements of the 1¼ co2 bands. *Journal of Quantitative Spectroscopy and Radiative Transfer*, 6(6):717–725, December 1966.

[18] B. Briggs. Design, fabrication, and characterization of tunable fabry-pérot optical filters with 470 nm central wavelength. Master's thesis, University of Kassel, 2010.

[19] W. Brunner and K. Junge. *Wissensspeicher Lasertechnik*. VEB Fachbuchverlag Leipzig, Leipzig, 3 edition, 1989.

[20] W. Brunner and H. Paul. Mode competition and its utilization in intracavity absorption spectroscopy. *Poznan Uniwersytet im Adama Mickiewicza Seria Fizyka*, 35:89–102, 1980.

[21] M. Carriero, M. Miorali, and C. Gommellini. Poisoning of lambda sensor: An experimental method to measure the lambda sensor switch velocity and its effect on air-fuel ratio excursion. October 1998.

[22] W. Chow and S. Koch. *Semiconductor-Laser Fundamentals: Physics of the Gain Materials*. Springer, 1 edition, August 1999.

[23] C. Cohen-Tannoudji. *Quantenmechanik*. Walter de Gruyter, 3 edition, May 2008.

[24] G. A. Coquin and K. W. Cheung. Electronically tunable external-cavity semiconductor laser. *Electronics Letters*, 24(10):599–600, May 1988.

[25] W. Demtröder. *Experimentalphysik 2: Elektrizität und Optik (Springer-Lehrbuch) (German Edition)*. Springer, 2 edition, 2002.

[26] W. Demtröder. *Experimentalphysik 3: Atome, Moleküle und Festkörper (Springer-Lehrbuch) (German Edition)*. Springer, 3., überarb. aufl. edition, April 2005.

[27] http://macro.lsu.edu/howto/solvents/dielectric constant.htm, January 2010.

Bibliography

[28] F. J. Duarte, editor. *Tunable Lasers Handbook (Optics and Photonics)*. Academic Press, December 1995.

[29] I. Duling. All-fiber ring soliton laser mode locked with a nonlinear mirror. *Opt. Lett.*, 16(8):539–541, April 1991.

[30] J. Eichler and H. J. Eichler. *Laser Bauformen, Strahlfuehrung, Anwendungen*. Springer, 6 edition, September 2006.

[31] J. Eschner. *Dynamik und Absorptionsempfindlichkeit eines Vielmoden-Lasers*. PhD thesis, University of Hamburg, 1993.

[32] A. Garnache, A. Kachanov, F. Stöckel, and R. Houdré. Diode-pumped broadband vertical-external-cavity surface-emitting semiconductor laser applied to high-sensitivity intracavity absorption spectroscopy. *J. Opt. Soc. Am. B*, 17(9):1589–1598, September 2000.

[33] J. Han, S. Baik, J. Jeong, K. Im, H. Moon, H. Noh, and D. Choi. Output characteristics of a simple fp-ld/fbg module. *Optics & Laser Technology*, 39(2):313–316, March 2007.

[34] T. Hansch, A. Schawlow, and P. E. Toschek. Ultrasensitive response of a cw dye laser to selective extinction. *Quantum Electronics, IEEE Journal of*, 8(10):802–804, October 1972.

[35] F. Harren, F. Bijnen, J. Reuss, L. Voesenek, and C. Blom. Sensitive intracavity photoacoustic measurements with a co2 waveguide laser. *Applied Physics B: Lasers and Optics*, 50(2):137–144, February 1990.

[36] E. Hecht. *Optik*. Oldenbourg Wissensch.Vlg, January 2005.

[37] www.heraeus.de.

[38] H. Hillmer. Sensorvorrichtung und verfahren zur ermittlung einer physikalischen größe. de 102004037519 b4, ep 000001771720 a1, wo 002006012825 a1. Patent, 2004.

[39] H. Hillmer. *Handbook of Lasers and Optics*, volume XXVI. Springer, 2007.

[40] E. D. Hinkley. High-resolution infrared spectroscopy with a tunable diode laser. *Applied Physics Letters*, 16(9):351–354, 1970.

[41] J. Hollas. *High Resolution Spectroscopy*. Wiley, 2 edition, October 1998.

[42] C. Huang, C. Cheng, Y. Su, and C. Lin. Mode competition in wide-range tunable dual-

Bibliography

wavelength semiconductor laser using nonidentical ingaasp quantum wells. In *Proceedings of 5th Pacific Rim Conference on Lasers and Electro-Optics*, page 55. IEEE, 2003.

[43] A. Imada, T. Mukai, A. Fujiwara, H. J. Lee, S. Hasegawa, and H. Asahi. Reduced temperature dependence of refractive index in tiingaas by addition of t1. In *16th IPRM. 2004 International Conference on Indium Phosphide and Related Materials, 2004.*, pages 465–468. IEEE, 2004.

[44] R. Jambunathan and J. Singh. Design studies for distributed bragg reflectors for short-cavity edge-emitting lasers. *Quantum Electronics, IEEE Journal of*, 33(7):1180–1189, August 2002.

[45] E. Kharasch. Every breath you take, we'll be watching you. *Anesthesiology*, 106(4):562–564, April 2007.

[46] K. Kieu and M. Mansuripur. All-fiber bidirectional passively mode-locked ring laser. *Opt. Lett.*, 33(1):64–66, January 2008.

[47] W. Köchner and M. Bass. *Solid-State Lasers: A Graduate Text*. Springer, 1 edition, May 2003.

[48] H. Krause. *Entwicklung eines Gassensors auf Basis der Modenkonkurrenz bei Halbleiterlasern*. PhD thesis, University of Kassel, 2010.

[49] H. Krause, J. Sonksen, J. Baumann, U. Troppenz, W. Rehbein, V. Viereck, and H. Hillmer. Rin spectra of a two-mode lasing two-section dfb laser for optical sensor application. In K. Panajotov, M. Sciamanna, A. Valle, and R. Michalzik, editors, *SPIE Photonics Europe 2008*, volume 6997, pages 22–31. SPIE, SPIE, May 2008.

[50] H. Kroemer. A proposed class of heterojunction injection lasers. *Proc. IEEE*, 51:1782+, 1963.

[51] R. T. Ku, E. D. Hinkley, and J. O. Sample. Long-path monitoring of atmospheric carbon monoxide with a tunable diode laser system. *Appl. Opt.*, 14(4):854–861, April 1975.

[52] E. Lacot, F. Stöckel, D. Romanini, and A. Kachanov. Spectrotemporal dynamics of a two-coupled-mode laser. *Physical Review A*, 57(5):4019–4025, May 1998.

[53] S. Larouche and L. Martinu. Openfilters: open-source software for the design, optimization, and synthesis of optical filters. *Appl. Opt.*, 47(13):C219–C230, May 2008.

[54] T. Lay. Fiber grating laser: a performance study on coupling efficiency of fiber microlens and the bragg reflectivity. *Optics Communications*, 233(1-3):89–96, March 2004.

Bibliography

[55] J. Liu. *Photonic Devices*. Cambridge University Press, May 2005.

[56] J. Lu, S. Fujita, T. Kawaharamura, H. Nishinaka, Y. Kamada, and T. Ohshima. Carrier concentration induced band-gap shift in al doped znmgo thin films. *Applied Physics Letters*, 89(26):262107+, 2006.

[57] H. Macleod. *Thin Film Optical Filters*. Taylor & Francis, 3 edition, January 2001.

[58] T. H. Maiman. Stimulated optical radiation in ruby. *Nature*, 187(4736):493–494, August 1960.

[59] P. Martina and R. Holdsworth. High-resolution infrared spectroscopy for in situ industrial process monitoring. *Spectroscopy Europe*, pages 7–15, October 2004.

[60] http://www.fiberoptic-solution.de/view/produkte/spleissgeraete/os_lid.htm, March 2010.

[61] http://www.orioninstruments.com/html/tools/dielectric.aspx, January 2010.

[62] B. A. Paldus, Jr, J. Martin, J. Xie, and R. N. Zare. Laser diode cavity ring-down spectroscopy using acousto-optic modulator stabilization. *Journal of Applied Physics*, 82(7):3199–3204, 1997.

[63] R. Paoletti, M. Meliga, G. Rossi, M. Scofet, and L. Tallone. 15-ghz modulation bandwidth, ultralow-chirp 1.55-μm directly modulated hybrid distributed bragg reflector (hdbr) laser source. *Photonics Technology Letters, IEEE*, 10(12):1691–1693, August 2002.

[64] I. Park, I. Fischer, and W. Elsässer. Highly nondegenerate four-wave mixing in a tunable dual-mode semiconductor laser. *Applied Physics Letters*, 84(25):5189–5191, 2004.

[65] A. Paruta. Solubility of several solutes as a function of the dielectric constant of sugar solutions. *Journal of Pharmaceutical Sciences*, 53(10):1252–1254, 1964.

[66] http://www.pharmawiki.ch, January 2010.

[67] A. Pötzl. Optimierung eines zwei - modigen faserbasierten laserresonators für sensoranwendungen. Master's thesis, University of Kassel, 2009.

[68] http://macro.lsu.edu/howto/solvents/refractive index.htm, January 2010.

[69] J. Reid, M. El-Sherbiny, B. K. Garside, and E. A. Ballik. Sensitivity limits of a tunable diode laser spectrometer, with application to the detection of no2 at the 100-ppt level. *Appl. Opt.*, 19(19):3349–3353, October 1980.

Bibliography

[70] J. Reid, J. Shewchun, B. K. Garside, and E. A. Ballik. High sensitivity pollution detection employing tunable diode lasers. *Appl. Opt.*, 17(2):300–307, January 1978.

[71] F. Rettig, R. Moos, and C. Plog. Poisoning of temperature independent resistive oxygen sensors by sulfur dioxide. *Journal of Electroceramics*, 13(1):733–738, July 2004.

[72] D. Romanini and K. Lehmann. Ring-down cavity absorption spectroscopy of the very weak hcn overtone bands with six, seven, and eight stretching quanta. *Journal of chemical physics*, 99(9):6287–6301, 1993.

[73] F. Römer. *Charakterisierung und Simulation optischer Eigenschaften von mikromechanisch abstimmbaren Filterbauelementen.* PhD thesis, University of Kassel, 2005.

[74] D. Rudan-Tasic and C. Klofutar. Characteristics of vegetable oils of some slovene manufacturers. *Acta Chim. Slov*, 46:511–521, 1999.

[75] V. Sankaranarayanan. Gas sensors market - an overview, August 2007.

[76] B. Scherer and J. Herbst. Pressure broadening of the oxygen a-band measured by laser absorption spectroscopy. volume 7222, pages 72220D+. SPIE, 2009.

[77] B. Scherer, W. Salzmann, J. Wöllenstein, and M. Weidemueller. Injection seeded single mode intra-cavity absorption spectroscopy. *Applied Physics B: Lasers and Optics*, 96(2):281–286, August 2009.

[78] A. Schirmacher. Test certificate ptb-4.51-4042164-09. Technical report, Physikalisch Technische Bundesanstalt, 38023 Braunschweig, July 2009.

[79] Michael A. Scobey. External cavity semiconductor laser with monolithic prism assembly, December 1998.

[80] V. Sherstnev, E. Grebenshchikova, A. Monakhov, A. Astakhova, N. Il'inskaya, G. Boissier, R. Teissier, A. Baranov, and Y. Yakovlev. A semiconductor whispering-gallery-mode laser with ring cavity operating at room temperature. *Technical Physics Letters*, 35(8):749–752, August 2009.

[81] K. Shiraishi, T. Chuzenji, and S. Kawakami. Polarization-independent in-line optical isolator with lens-free configuration. *Lightwave Technology, Journal of*, 10(12):1839–1842, August 2002.

[82] J. C. Shrestha. Characterization of fiber based laser by means of relative intensity noise (rin). Master's thesis, University of Kassel, 2010.

Bibliography

[83] S. Shugan. Price-quality relationships. *Advances in Consumer Research*, XI:627–632, 1984.

[84] J. Sierks, V. M. Baev, and P. E. Toschek. Enhancement of the sensitivity of a multimode dye laser to intracavity absorption. *Optics Communications*, 96:81–86, 1993.

[85] http://www.sigmaaldrich.com, January 2010.

[86] M. Sigrist, editor. *Air Monitoring by Spectroscopic Techniques*. John Wiley & Sons, 1 edition, March 1994.

[87] Brian C. Smith. *Fundamentals of Fourier Transform Infrared Spectroscopy*. CRC Press, 1 edition, December 1995.

[88] S. A. Sochinaz. Propofol datasheet 91/155 ewg. March 2006.

[89] J. Sonksen, M. Ahmad, N. Storch, H. Krause, S. Blom, A. Pötzl, and H. Hillmer. Controlling and tuning the emission of semiconductor optical amplifier for sensor application by means of fiber bragg gratings. In *Proceedings of the 8th WSEAS international conference on Microelectronics, nanoelectronics, optoelectronics*, pages 59–62, Stevens Point, Wisconsin, USA, 2009. WSEAS.

[90] Refined soybean oil u.s.p. - general specifications. Technical report, 7 Avenue L, Newark, NJ 07105, January 2005.

[91] http://www.laborpraxis.de/index.cfm?pid=9275&title=absorptionsspektroskopie, April 2010.

[92] J. Stone. Reduction of oh absorption in optical fibers by oh - od isotope exchange. *Industrial & Engineering Chemistry Product Research and Development*, 25(4):609–621, December 1986.

[93] O. Svelto. *Principles of Lasers*. Springer, 4 edition, September 2009.

[94] A. Takamizawa, G. Yonezawa, H. Kosaka, and K. Edamatsu. Littrow-type external-cavity diode laser with a triangular prism for suppression of the lateral shift of output beam. *Review of Scientific Instruments*, 77(4):46102+, 2006.

[95] A. Thony. New developments in co2-laser photoacoustic monitoring of trace gases. *Infrared Physics & Technology*, 36(2):585–615, February 1995.

[96] P. E. Toschek and V. M. Baev. *One is not enough: intra-cavity spectroscopy with multimode lasers*, volume 54 of *Optical Sciences*, pages 89–111. Springer Verlag, Berlin, 1987.

Bibliography

[97] R. Toth. Measurements of h2 16o line frequencies and strengths: 11610 to 12861 cm-1. *J. Molec. Spectr.*, 166:176–183, 1994.

[98] B. Van Zeghbroeck. *Principles of Semiconductor Devices*. University of Colorado, 2007.

[99] C. Voumard, R. Salathé, and H. Weber. Mode selection by etalons in external diode laser cavities. *Applied Physics A: Materials Science & Processing*, 7(2):123–126, June 1975.

[100] G. Wedler. *Lehrbuch Der Physikalischen Chemie*. Vch Verlagsgesellschaft mbH, 4 edition, 1997.

[101] J. White. Long optical paths of large aperture. *J. Opt. Soc. Am.*, 32(5):285, May 1942.

[102] A. Wicht, M. Rudolf, P. Huke, R. H. Rinkleff, and K. Danzmann. Grating enhanced external cavity diode laser. *Applied Physics B: Lasers and Optics*, 78(2):137–144, January 2004.

[103] G. Wysocki, R. F. Curl, F. K. Tittel, R. Maulini, J. M. Bulliard, and J. Faist. Widely tunable mode-hop free external cavity quantum cascade laser for high resolution spectroscopic applications. *Applied Physics B: Lasers and Optics*, 81(6):769–777, October 2005.

[104] K. Yoshino and A. Blankstein. Lethal levels of anesthetic propofol killed michael jackson. August 2009.

[105] L. Yu-Lung and C. Han-Sheng. Measurement of thermal expansion coefficients using an in-fibre bragg-grating sensor. *Measurement Science and Technology*, 9(9):1543–1547, 1998.

[106] P. Zorabedian and W. R. Trutna. Interference-filter-tuned, alignment-stabilized, semiconductor external-cavity laser. *Opt. Lett.*, 13(10):826–828, October 1988.

List of Figures

1.1. Organigram of the partners involved in the registration and exploitation of patent DE 10 2004 037 519 B4. 6

2.1. Extremely simplified ICAS laser [38]. 9

2.2. Working principle of a laser, consisting of pumped active material between two mirrors: the pump energy applied to the active material establishes inversion, which provides amplification by stimulated emission of radiation. 10

2.3. Energy schemes of active material
(a) three level system, valid for ruby lasers [58]
(b) a four level system, valid for solid state lasers and semiconductor lasers. . . . 11

2.4. Wavelengths supported by the FP resonator
(a) Visualization of constructive interference of the fundamental and first order mode in the FP resonator
(b) Normalized output of a FP Resonator calculated by the transfer function given in eq. 2.8 for different mirror reflectivities. 14

2.5. Illustration of the FWHM of a normalized Gaussian function. 16

2.6. (a) Threshold condition for laser modes.
(b) Measurement of the output intensity of a commercial laser diode as a function of injection current for different temperatures. The kink indicates threshold current. 17

2.7. (a) Schematic cross section of a FP semiconductor laser diode with electrical contacts.
(b) Example of the optical output spectrum of Thorlabs S1FC1550 FP laser source, provided by the vendor. 19

2.8. (a) Band diagram of a PIN homo junction [22],
(b) Band diagram of the forward biased double hetero structure PIN junction [39]. 21

135

List of Figures

2.9. (a) Measurement of the red shift of the output wavelength of a commercial semiconductor laser under increasing temperature.
(b) Measurement of the blue shift of the output wavelength of a commercial semiconductor laser under increasing injection current. 23

2.10. (a) Schematic cross section of a FP semiconductor laser diode with DBR mirrors.
(b) Calculated spectral reflectivity of a DBR consisting of 23.5 periods of $\lambda/4$-layers SiO_2/SiN. 24

2.11. Example of an external cavity semiconductor laser with a grating as the wavelength selective element (Littrow configuration) [30]. 25

2.12. The generation of an oscillating dipole moment by rotation of an electric dipole. 28

2.13. Vibration modes of a molecule consisting of three atoms
(a) In the symmetric case the vibration does not cause a change in dipole moment and no interaction of EM-radiation is possible.
(b), (c) Asymmetric modes cause a change in dipole moment and enable interaction with EM-radiation. 29

2.14. Eigenvalues of the Morse potential in comparison to those of the harmonic potential. Spacing is reduced for the asymmetric Morse potential and converges to 0 at the dissociation energy of the molecule. 30

2.15. Rotational-vibrational absorption spectrum of HCn, calculated from HITRAN database The P- and R-branch are clearly observed, the Q-branch is not observed because it corresponds to a forbidden transition. 32

2.16. (a) Schematic setup of a conventional absorption experiment: a laser is used as light source and its emission subjected to absorption outside of the resonator after amplification.
(b) Schematic setup of an ICAS experiment: high sensitivity is achieved by placing the absorber inside the resonator. 37

2.17. Multi mode ICAS vs. single mode ICAS
(a) In the highly sensitive multi mode case complex and expensive dye or fiber lasers and high resolution measurement equipment is required.
(b) In the single mode case much more cost efficient devices like diode lasers and photo detectors can be applied, but sensitivity is significantly lower. 43

3.1. Starting point of this work: the laser system from [48]. 47

List of Figures

3.2. Optical output measured for the laser system developed in [48]. M_i and M_o are found at 1541.98 nm and 1542.47 nm with FWHM of 0.08 ± 0.01 nm and (0.06 ± 0.01) nm, respectively. Mode intensity differs by 3.6 dB. Side modes are found at 1541.53 nm and 1543 nm. 49

3.3. FBG tuning mechanism
(a) The FBG (red) is glued onto an Aluminum block (black). Heating the block causes a shift of its reflectivity to longer wavelengths due to thermal expansion. (b) Measured laser output under FBG tuning between 33°C and 65°C. 50

3.4. Measured wavelength stability of the laser system documented in [89]. Fluctuation of 0.01 nm are observed after stabilization. 51

3.5. (a) Laser system [48] with one fixed ratio coupler.
(b) Implementation of two variable couplers: V_i enables tailoring of threshold current and V_o equalization of mode intensities. 53

3.6. Schematic of the setup used for the characterization of variable couplers $V_{i,o}$. . . 55

3.7. (a) Measurements of coupling ratio of the variable couplers as a function of knob setting: the red curve denotes coupler V_i, blue curve coupler V_o.
(b) Specification provided by the vendor. 56

3.8. Schematics of the collimator arrangement implemented. 58

3.9. Schematics of the setup used for the characterization of the collimator arrangement implemented. 60

3.10. Possible positions A-D for the implementation of the collimator arrangement. . . 60

3.11. Laser system including the collimator arrangement in position B_1. 61

3.12. Stabilization behavior of the FBG feedback system in the initial laser system by [48]. Final stability is ultimately limited by thermal fluctuation of the environment. 63

3.13. Measured spectral output of the SOA. A superstructure is observed on the broad Gaussian gain curve. 66

3.14. Schematic illustration of the working principle of a polarization dependent optical isolator. 67

3.15. Schematics of the setup used for the characterization of optical isolators. 68

3.16. Measured spectral throughput of the isolators measured in reverse direction. . . 68

3.17. Laser system including isolators, numbered according to tab. 3.4. 69

3.18. Schematics of the setup used for the measurement of splicing loss. 70

3.19. Laser system including all refined splices a-j. 71

List of Figures

3.20. (a) Broad band transmission spectrum and
(b) Narrow band transmission spectrum of 10 mm Suprasil provided by the vendor [37]. 73

3.21. (a) Sample cells purchased.
(b) Simulation of the 10 mm sample cell's spectral transmission. 74

3.22. (a) Complete sample cell holder.
(b) Schematics of the optical path through the holder including sample cell and beam offset compensation. 76

3.23. (a) Enhanced laser system developed in this work in comparison to
(b) the laser system from [48]. 79

4.1. The laser system developed in this work. 81

4.2. (a) Measurement of the broad band output of the enhanced laser system at $I_{inj} = 113.6$ mA and $\Theta_{SOA} = 25°C$.
(b) Detail of the two individual laser modes at $\lambda_i = 1542$ nm and $\lambda_o = 1542.5$ nm. 82

4.3. Measurements on SOA gain curve at $\Theta_{SOA} = 15°C$ for injection currents from $I_{inj} = 5 - 150$ mA. 83

4.4. (a) Measurements on the center wavelength of the gain curve of the SOA at $\Theta_{SOA} = 15, 20, 25$ and $30°C$ for injection currents from $I_{inj} = 5 - 150$ mA.
(b) Measurements on the FWHM of the gain curve of the SOA at $\Theta_{SOA} = 15, 20, 25$ and $30°C$ for injection currents from $I_{inj} = 5 - 150$ mA. 84

4.5. Measurements on the SOA gain curve at $I_{inj} = 80$ mA for SOA temperatures from $\Theta_{SOA} = 10 - 30°C$. 85

4.6. (a) Measurements on the center wavelength of the gain curve of the SOA at $I_{inj} = 80, 100, 120$ and 140 mA for SOA temperatures from $\Theta_{SOA} = 10 - 30°C$.
(b) Measurements on the FWHM of the gain curve of the SOA at $I_{inj} = 80, 100, 120$ and 140 mA for SOA temperatures from $\Theta_{SOA} = 10 - 30°C$. 86

4.7. Schematics of the setup used for the measurements of threshold current on mode M_i. 87

4.8. (a) MEasured coupling ratio for coupler V_i between in- and tp-port.
(b) Calculated loss for 2 passes (in-tp, tp-in) on coupler V_i. 88

4.9. Exponential approximation of threshold current as a function of the 2 pass loss shown in fig. 4.8 b. 89

4.10. Measured FWHM of M_i and M_o at SOA temperature of $\Theta_{SOA} = 15°C$. The drastic drop at 65 mA corresponds to laser threshold. 90

List of Figures

4.11. SNR measured as a function of injection current for M_i and M_o at SOA temperatures of 15, 20 and 25°C. 92

4.12. Comparative measurement of the output spectra of the laser system from [48] and the one developed in this work at $\Theta_o = 33$ and 65°C. 93

4.13. Full data set of the thermal tuning response measurement
(a) on the laser system developed in this work
(b) on the laser system from [48]. 94

4.14. (a) Measured intensity difference $\Delta \mathscr{I} = \mathscr{I}_o - \mathscr{I}_i$ under thermal tuning for both laser systems.
(b) Measurement on the tuning response of the laser system by [48] and the one developed in this work. 95

4.15. (a) Stabilization behavior of the FBG feedback system in the laser system developed in this work, where final stability is determined by long term drift, in comparison to
(b) the laser system from [48], where final stability is determined by thermal fluctuation of the environment. 96

4.16. Structure of Propofol. 102

4.17. (a) Broad band absorption coefficient calculated from own transmission experiments on Propofol.
(b) Calibrated absorption coefficient provided by the PTB Braunschweig between 1500 and 1600 nm. 105

4.18. Schematic setup of Perkin Elmer Lambda 900 spectrophotometer. 106

4.19. (a) Broad band absorption coefficient calculated from own transmission experiments on the solvents considered for the POC.
(b) Detail of the absorption coefficient of the solvents considered for the POC along with the calibrated absorption coefficient of Propofol provided by the PTB Braunschweig. 107

5.1. Schematics of the setup used for the POC measurements. 109

5.2. Normalized output spectra of the laser system with absorption from Propofol
(a) in 1 ml DCM at $\Delta\lambda = 1$ nm (series 5) and
(b) in 1 ml DCM at $\Delta\lambda = 0.5$ nm (series 6). 113

5.3. Comparison of normalized output spectra of the laser sytem with absorption from Propofol
(a) in 1 ml DCM at at $\Delta\lambda = 1$ nm (series 5) and
(b) in 2 ml DCM at at $\Delta\lambda = 1$ nm (series 7). 115

List of Figures

5.4. Comparison between Mode Intensity Ratio for $\mathscr{I}_{int}$ and $\mathscr{I}_{max}$ for Propofol in 1 ml DCM at $\Delta\lambda = 1$ nm (series 5) and in 2 ml DCM at $\Delta\lambda = 1$ nm (series 7). . 116

5.5. Comparison between different concentration definitions
   (a) Molarity M,
   (b) Molar Fraction χ and
   (c) parts per million 'ppm'. 118

5.6. Comparison of Mode Intensity Ratio Υ versus concentration for measurement series 1-9. 119

5.7. Comparison of Mode Intensity Ratio Υ versus ΔA for measurement series 1-9. . 121

5.8. Comparison of Mode Intensity Ratio Υ versus ΔA for measurement series 6. . . . 122

A.1. Schematics of the setup used for the POC measurements. 145

A. Appendix

Datasheet SOA

A. Appendix

INPHENIX

1550nm SOA Test Data for Heinrich

SOA can amplifies the signal if the wavelength of input signal is within 3dB or 6 dB bandwidth range of the ASE spectrum of SOA, see figure 1 (a) and (b). However the gain would not same for different wavelengths, see the gain vs. wavelength profile in figure 2. This SOA should not problem to reach the specifications that Kassel University requested for 1545nm and 1585.4nm signals. For small signal:

    Gain> 20dB
    FMHW>30nm
    Psat>8dBm
    PDL<1dB
    NF~9dB which is higher than 7dB.

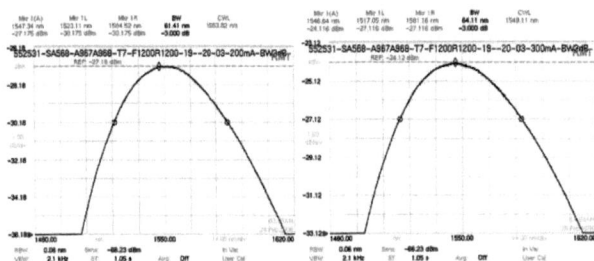

Fig. 1. ASE spectrum of 1550nm SOA (a) at 200mA (b) at 300mA

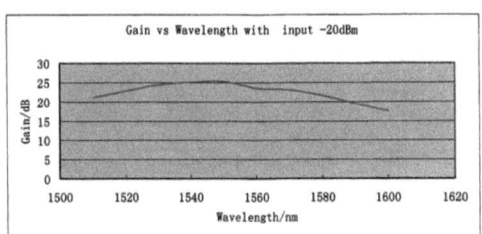

Fig. 2. The gain profile vs. wavelength with input signal of −20dBm

2/14/2008

Datasheet FBG

```
|===================================================|
|             Fiber Bragg Grating Control Sheet     |
|===================================================|
| Company              | Advanced Optics Solutions GmbH |
|                      | Ammonstr. 35                   |
|                      | D-01067 Dresden, Germany       |
|                      |                                |
| Phone                | +49 (0)351 4960 193            |
| Fax                  | +49 (0)351 4960 194            |
| E-mail               | info@aos-fiber.com             |
| WWW                  | www.aos-fiber.com              |
|                      |                                |
| Date                 | 02.06.08                       |
|                      |                                |
| SerNo.               | 02060848                       |
|---------------------------------------------------|

Notes:


|===================================================|
|                 Measured Parameters               |
|===================================================|
| Peak-Wavelength    (nm)   | 1542.385               |
| Peak-Frequency     (THz)  | 194.369                |
| Center-Wavelength  (nm)   | 1542.422               |
| Center-Frequency   (THz)  | 194.364                |
|                           |                        |
| Transmission Loss  (dB)   | 21.0                   |
| Reflectivity       (%)    | 99.2                   |
| SNR (>dB) @Peak+/-1nm     | 25.5                   |
|---------------------------+------------+-----------|
| Bandwidth                 | nm         | GHz       |
|---------------------------+------------+-----------|
| 1dB                       | 0.081      | 10.175    |
| 3dB                       | 0.168      | 21.133    |
| 20dB                      | 0.385      | 48.525    |
|===================================================|
```

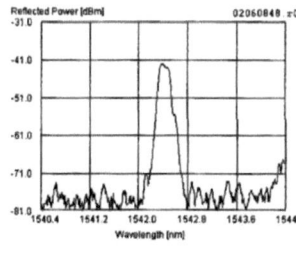
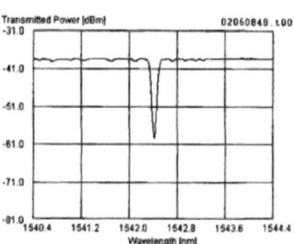

A. Appendix

Test Certificate Fixed Coupler

2x2 Coupler Test Report

Invoice No.: HPGW08085

Product Code	WFSC-X-P-13-15-9-C-5-10-FA
Serial Number	08020693
Fiber Type	Corning SMF-28e
Operation Wavelength (nm)	1550±40
Directivity (dB)	≥ 55.0

In-port	Out-port	Insertion Loss (dB)	Test CR (%)	REQ CR (%)	PDL (dB)	Excess Loss (dB)
Blue	Red	0.70	87.09	87	0.01	0.10
	None	8.99	12.91	13	0.02	
None	Red	8.97	12.98	13	0.01	0.10
	None	0.71	87.02	87	0.01	

Note: Exclude insertion loss of connectors
In/Out connector type: __FC/APC-FC/APC__ (IL≤0.20dB/Per connection)
Tested By: __F-55__ Checked By:

Setup and Measurement Overview

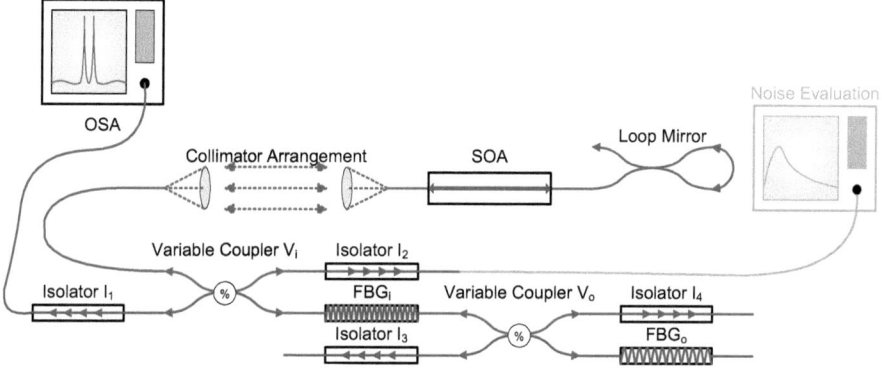

Figure A.1.: Schematics of the setup used for the POC measurements.

A. Appendix

Index	Solvent	Volume/Cell Size	SOA Current	$\Delta\lambda$
1 —▲—	Acetone	1 ml / 5 mm	125 mA	1 nm
2 ---▲---				0.5 nm
3 —■—		2 ml / 10 mm		1 nm
4 ---■---				0.5 nm
5 —▲—	DCM	1 ml / 5 mm	125 mA	1 nm
6 ---▲---				0.5 nm
7 —■—		2 ml / 10 mm	160 mA	1 nm
8 ---■---				0.5 nm
9 —△—	Acetone	0.75 ml / 5 mm	110 mA	1 nm
10	Soy Oil	1 ml / 5 mm	125 mA	1 nm
11	2-Propanol	— —	— —	— —

Table A.1.: Overview of the measurement taken for the POC.

Acknowledgment

Completing a Ph.D in only 2 years and 9 months is an achievement that I could never have accomplished on my own. Here, I would like to acknowledge the support of everybody, who directly or indirectly contributed to the success of this work.

First of all, I would like to thank Prof. Dr. H. Hillmer for giving me the opportunity to carry out the research documented in this thesis at INA. The extraordinary trust he put in to me and his staff every day and his enthusiasm and drive have been a great source of motivation throughout the time I spent with this project. I also thank Prof. Dr. J. Börcsök, Prof. Dr. A. Bangert and Prof. Dr. K. Röll for co-supervising my thesis.

The NanoNose-group that formed during my time I would like to thank for great team game and many inspiring discussions. This work demonstrates what we have achieved together!

In the design and fabrication of the many custom made components created in this work the technical staff and workshops have been very helpful. Especially "The Man with the Golden Hands" , Dietmar Gutermuth, has earned my deep respect for his outstanding practical skills!

For providing samples I would like to thank Dr. H.-O. Maier from B.Braun Melsungen AG. For providing an "icy atmosphere" my thanks are directed at the cell Biology group of Prof. Dr. Maniak. The group of organic Chemistry I would like to thank for help with spectroscopic measurements.

I would like to express my gratitude to Mr. Köhler, Mr. Wittzack and Mr. Jäkel for being great wingmen and outstanding friends and giving me the opportunity to do the same for them.

The rest of the colleagues at INA I would like to thank for providing a creative, helpful and supportive working environment that I surely will miss.

Last but not least I want to acknowledge the unconditional support from my mother in all the years leading up to this moment.

Die VDM Verlagsservicegesellschaft sucht für wissenschaftliche Verlage abgeschlossene und herausragende

Dissertationen, Habilitationen, Diplomarbeiten, Master Theses, Magisterarbeiten usw.

für die kostenlose Publikation als Fachbuch.

Sie verfügen über eine Arbeit, die hohen inhaltlichen und formalen Ansprüchen genügt, und haben Interesse an einer honorarvergüteten Publikation?

Dann senden Sie bitte erste Informationen über sich und Ihre Arbeit per Email an *info@vdm-vsg.de*.

Sie erhalten kurzfristig unser Feedback!

VDM Verlagsservicegesellschaft mbH
Dudweiler Landstr. 99   Telefon +49 681 3720 174
D - 66123 Saarbrücken   Fax   +49 681 3720 1749
www.vdm-vsg.de

Die VDM Verlagsservicegesellschaft mbH vertritt

Printed by Books on Demand GmbH, Norderstedt / Germany